CHIL

'An extraordinary and painstakingly researched history of the Nazis' most bizarre racial experiment.'
Daily Express

'Every thinking person today should read the book because it shows only too clearly the terrible consequences of imposing any form of sexual control on society.'
Nursing Mirror

'A unique glimpse into a brave new world gone mad.'
Library Journal

'Even the most habituated student of Nazism will be appalled by this account of the Lebensborn 'Fount of Life' organization, which not only ran SS breeding centres but participated in mass kidnappings of children in occupied territory.'
Kirkus Reviews

Clarissa Henry
& Marc Hillel

Children of the SS

Translated from the French by Eric Mosbacher

CORGI BOOKS
A DIVISION OF TRANSWORLD PUBLISHERS LTD

CHILDREN OF THE SS

A CORGI BOOK 0 552 10614 3

Originally published in Great Britain by Hutchinson & Co (Publishers) Ltd.
Published in French as AU NOM DE LA RACE 1975

PRINTING HISTORY
Hutchinson edition published 1976
Corgi edition published 1977

This book is set in Intertype Times

Corgi Books are published by
Transworld Publishers Ltd,
Century House, 61–63 Uxbridge Road,
Ealing, London W5 5SA
Made and printed in Great Britain by
Cox & Wyman Ltd., London, Reading and Fakenham

Contents

Illustrations

Between pages 176 and 177

PART ONE

The Orphans of Shame

'The most important things are not
always to be found in the files.'

Goethe

I. The Trail

We spent nearly three years tracking down every available source about the Lebensborn organization. We had heard of the human stud-farms created by Heinrich Himmler during the Second World War; we knew that they were called Lebensborns. The term Lebensborn was a Nazi neologism composed of *Leben*, 'life', and the medieval *Born*, 'fount', 'spring' or 'source': *Lebensborn*, fountain of life. The subject has been mentioned in history books, but never dealt with thoroughly.

To some writers the Lebensborns were maternity homes for unmarried mothers, to others they were brothels for the exclusive use of the SS. At all events they seem to agree that the real aim of these institutions was the breeding of a Nordic super-race with the aid of men and women carefully selected in accordance with the racial principles of the Third Reich. What had been the outcome of this enterprise, and after the war what had become of all the children born as a result of it?

We did not know how to set about our inquiry. After reading the literature on the SS we set about searching the relevant files. We soon discovered that the Lebensborn organization was not confined to Germany. We drew up a list of countries in which they had been established, and wrote to the appropriate Ministries in each, asking whether it might be possible to study any available documents. All the replies were similar and all were negative. Surprise ('What business is it of yours?') was often associated with disapproval ('It is none of your business') and even condemnation ('It is shocking to want to expose these children to such shame now').

Official statistics about the Lebensborn children are nowhere to be found. Their past, like their origin, has been wiped out; born of unknown parents, officially they do not exist.

Faced with all these difficulties, we might perhaps have abandoned our quest if one day in November 1971 Clarissa had not come home from working on documents in Paris with a discovery that moved her deeply. She had discovered that the provision of special maternity homes for turning out Nordics was only part of the Lebensborn activities. All over Europe the Germans had kidnapped thousands of 'racially valuable' children, taken them from their families to Germanize them. That was one of the ways of helping the super-race to be fruitful and multiply.

This gave a more poignant dimension to our inquiry. The curiosity by which we had been previously motivated gave way to a deeper feeling. We thought of the parents robbed of their children, of the fate of the young Poles, Czechs, Yugoslavs and Russians kidnapped and plunged into an unknown world by the SS. There were over 200000 taken from Poland alone.

We tracked down the organization charts of the Lebensborn organization and traced the careers of its principal personalities. Two of these attracted our special attention. Max Sollmann, the administrative head of the organization, was born at Bayreuth in 1904, joined the Nazi Party in 1922 (member no. 14528), and for unknown reasons had lived as an expatriate in Colombia until 1934. He joined the SS in 1937 (member no. 903), and reached the rank of Standartenführer (corresponding to that of colonel). In March 1939 he was decorated with the Blood Order for his part in the Munich *Putsch* of 1923, in which he had been one of the youngest participants. The medical chief of the organization was Gregor Ebner, born in Bavaria on 24 June 1892 and a member of the party from 1930. He joined the SS (member no. 1257), and became a specialist lecturer on the 'problems of racial selection'. He reached the rank of Oberführer in the SS (corresponding to that of brigadier-general) and was a close friend of Himmler's.

We were able to follow the careers of these men and their aides up to 1945 (we carefully studied the reports of the Lebensborn trial in 1947–8), but up to that year only. In common with all German citizens, the privacy of former members of the SS, including war criminals, is protected by law. We consequently abandoned all hope of tracking down the principal personalities of the organization. We presumed that Gregor Ebner was dead and that Max Sollmann had gone back to South America, and so had to fall back on other witnesses of lesser importance – nurses, doctors, unmarried mothers and children born in the Lebensborns. But here, too, we came up against a wall of silence.

We picked up the trail again unexpectedly. While consulting files in London Clarissa came across a 1955 press cutting in which a nurse named Paula was mentioned who had worked in a Lebensborn at Steinhöring, near Munich. After the war it had been turned into a home for handicapped children, and the nurse had continued working there. Here at last was a lead, the only one we had.

Steinhöring is a small village, like many others in Bavaria. The home is a big building on the outskirts. Fields stretch as far as the eye can see.

We asked a woman in the courtyard whether we could see Paula.

'Paula who?' she said.

'You know . . . she worked here during the war.'

'Oh, Paula Hessler. She retired ten years ago, but she lives in a little house quite near here.'

It was then that luck intervened in an extraordinary fashion. We decided to look up Paula Hessler in the local telephone directory. But Clarissa picked up the Munich directory by mistake. She turned to the letter S:

'Look, Max Sollmann,' she said. 'Supposing that's our man?'

'It can't be, it's impossible.'

'Why not ring him up and see?'

After a long discussion in Munich, we decided to try. If by any chance the man turned out to be 'our' Max Sollmann, Clarissa would pass herself off as a young Frenchwoman named Ingrid, born in a Lebensborn, with whose file we were familiar.

It was carnival time in Munich. We chose a telephone box in the centre of the city, and Clarissa dialled the number. A male voice replied.

'Herr Sollmann?'

'Speaking.'

'Herr Max Sollmann?'

'Yes.'

'Herr Max Sollmann of the Lebensborn?'

There was a moment's hesitation.

'Yes.'

Clarissa was suddenly afraid. Trembling and sincerely moved, she said:

'My name's Ingrid. I was born in a Lebensborn near Paris, at Lamorlaye. I've heard such a lot of things, I want to find out the truth about the place.'

'Where are you?' Max Sollmann said without hesitation.

'Near the central post office.'

'Come and see me.'

'When?'

'Right away.'

We jumped into a taxi. The unassuming building in which Max Sollmann lives is in the suburbs of the city. When Clarissa announced herself on the entryphone, the front door opened immediately, and in we went. Sollmann's flat is on the top floor, the fourth. While we climbed the staircase he watched us, leaning over the banister. He was an imposing figure, and he remained perfectly motionless. We could not help visualizing him in his SS uniform. His whole attitude, his impassibility, the gaze that he fixed on us, reminded us of a sentry in the watch-tower of a concentration camp checking on the inmates at the entrance to the camp. Clarissa was frightened; I concentrated on our task. At last we were meeting a man whose actions we had studied in books and files for more than two years.

Sollmann looked at me mistrustfully, without speaking; I explained that I was Ingrid's husband, that I was a journalist, and that I intended to write a book about my wife's childhood.

Max Sollmann was seventy, but did not look it. In spite of the constant trembling that shook his frame, he gave an impression of great physical strength. Whenever his blue eyes fell on us, the old man disappeared and gave way to the head of the Lebensborn organization, the man the nurses had called 'handsome Max'.

'You're in luck,' he said. 'My wife has gone out to see the carnival. If she had been at home I should not have let you in, because I promised her never again to talk to anyone about the Lebensborns.'

The flat was simply furnished, but it was bright and comfortable. On the bookshelves a leather-bound copy of *Mein Kampf* stood cheek by jowl with the Bible. Sollmann produced a bottle of red wine, offered us some, and went on drinking throughout the interview.

As soon as he started talking about the Lebensborns, the documents faded from our minds. What he told us was alive. 'Reichsführer,' he had said to his close friend Himmler, 'why don't we publish some articles in *Das Schwarze Korps* [the SS journal] and put a stop to all these rumours about the Lebensborns?' 'My dear Sollmann,' the Reichsführer had replied, 'my dear Sollmann, our conscience is clear, we have no need to justify ourselves.'

'Put this in your book, sir,' Sollmann told me. 'No one has written it yet. I can no longer do anything, because of my work and the state of my health. I'm no longer a fighter. It's up to the younger generation. Perhaps one day they will recognize that the Lebensborns were a good thing, and people will look back with regret at the opportunities that were wasted.'

I asked him whether he had known about the concentration camps.

'No,' he replied, 'practically no one did.'

'And you expect me to believe you?'

'You must. You have my word of honour.'

He went on to talk at length about his only child, for whose death the Americans were responsible.

'My wife had to undergo four operations before she could conceive. The baby was born in 1945, a few days after she was arrested by the Americans, heaven knows why. In prison she received no proper attention, and the child died. And do you know what the Americans did? They laid the dead baby on my wife's belly. Well, what do you say to that? What an act of barbarism. The two of us have never got over it. Just imagine it, our only child. We paid dearly for our ideals. And everything we did was done in the name of charity, in the name of mercy.'

On the subject of the Lebensborn organization his voice was loud and firm. 'All the stories you hear about it are false,' he said. 'They are lies made up by people who keep their own conscience clean by putting all the blame on us. All the Germans marched with Hitler, all of them, believe me. But we are the outcasts, we of the SS. It makes things so much simpler.'

The most extraordinary thing was his concern and solicitude for Clarissa. Though the comparison may seem shocking, his attitude towards my wife reminded us of the tenderness and concern that must have been shown by American Jews, for instance, to child survivors of the concentration camps. Believing her to be Ingrid, he took Clarissa under his wing immediately, and there was nothing he was not prepared to do for her. 'Would you like me to get you a certificate stating you were born in a Lebensborn?' he said. He spoke to her with a great deal of gentleness, and wanted to know all about her life. Had she been well treated? Was she happy? To me he spoke in a different tone. I was an outsider, a stranger, but to him Clarissa was still one of them. She belonged. Former members of the SS are generally highly reticent, and it was for her sake, and her sake only, that he broke his silence.

There was also something incredible about our meeting with SS Oberführer Gregor Ebner, who at the university had belonged to the same students' fraternity as Heinrich Himmler. We believed he must be dead, and Sollmann thought so too. But a young German doctor claimed to have met him in 1963 in the lake district near Munich. Was this true, and, if so, could he still be alive? Encouraged by the Sollmann precedent, we began by searching the telephone directories, but there was no trace of a Gregor Ebner. Having learned by experience that no help was to be expected from official sources, we approached a German colleague in the Munich area. He soon found a Dr Ebner. But he was only thirty years old. Our friend wondered whether he might not be the son of the former SS general, but we knew that 'our' Ebner had three daughters. Once more chance intervened. One day our friend's young son, aged about ten, was riding a bicycle in the neighbourhood when a tyre burst. He propped his bicycle against the wall of a house, glanced at the names on the letter-boxes, and noticed that one of them was Gregor Ebner. When he got home he announced triumphantly: 'If your French friends are still looking for Gregor Ebner, I've found him.'

This Gregor Ebner lived in the village of Wolfratshausen, not far from the Starnbergersee. We prowled around the house for a whole week, afraid that our French number-plate would be noticed. In the course of over two years we had come to know so much about him that we felt he must know we were on his trail.

After a great deal of hesitation I ended by ringing the bell. A fair-haired old lady appeared at a third-floor window.

'We come from Steinhöring and should like to talk to you,' I said.

'Come up,' she replied.

Frau Ebner opened the door. The flat was more modest than Sollmann's. Our first sight of Ebner was of a bedridden old man lying helpless on a couch with a rug over his knees. Some sort of rubber toy lay within reach of his hand. His face was emaciated and ill-shaven. Every so often a kind of rattle came from his twisted mouth. But his lively, very blue eyes seemed to confirm his wife's assurance that he understood everything.

For a moment we were speechless, unable to take in that this helpless old man had been the chief medical officer of the Lebensborn organization and had personally ordered the sterilization of many children and supervised the selection and Germanization of thousands of kidnapped children in the occupied territories. As soon as we got over the shock, I explained to Frau Ebner that Ingrid, my wife, had been born in a Lebensborn and wanted to find out the truth about those homes about which so many strange stories were told.

This piece of information had an extraordinary effect on Frau Ebner. Like Max Sollmann, believing Ingrid to be a Lebensborn child, she started talking volubly about the institutions of which her husband had been in charge.

'Why were fair, blue-eyed women chosen for the Lebensborns?' I asked.

'It was the fashion. You know how these things happen. There's a fashionable type of woman today too. Then the emphasis was on the Nordic type, I mean tall and blonde, it was the fashion.'

'And what is one to make of the reputation these places had of being brothels or human stud-farms?'

'It's disgraceful that such things should be said about them. The Lebensborns are denigrated solely because they were organized by the SS. After all, how many illegitimate children were born in Germany of American and Negro fathers after the war? The mothers would have been delighted if such remarkable institutions had been available for them to have their children in in peace and comfort. Now that the Negroes and the Americans have gone, they're left with the children, without any maintenance allowance or anything at all. In the Lebensborns at least the mothers were helped and the children were looked after.'

'But not any mother. If a girl of Jewish or gypsy or Slav origin had applied, would she have been accepted?'

'Yes, of course, everyone was helped.'

'And what happened to the foreign children, what became of them?'

'I don't know, but at all events they were well treated.'

You knew Heinrich Himmler. One hears such a lot of things about him. What sort of man was he?'

A gleam came into Frau Ebner's blue eyes. I looked at her husban, who was clumsily fingering his rubber toy. He did not react. We felt he was silently taking part in the conversation. He had been paralysed since a bad fall on the pavement during the previous year, but his brain was intact.

'Herr Himmler and my husband were old friends. They met in a students' association when they were both students in Munich . . . He was a pleasant man, he often came to see us before the war. Sometimes he would bring his daughter Gudrun, who played with our children. He was a good, honest, helpful man. Unfortunately, during the war we saw very little of him, because he was kept so busy by all those inmates in the camps.'

When we reminded her that in 1945 Himmler committed suicide in a British prisoner-of-war camp to avoid facing his judges, Frau Ebner could not repress a sigh.

'*Mein Gott*, how unfair it all is!' she exclaimed.

More striking than the strange things she said was her solicitude towards my wife. She was sorry to learn that 'Ingrid' had suffered at not having a family, and she behaved towards her like a mother. 'Oh, my poor child, it must have been dreadful,' she said. Suddenly she took Clarissa by the hand and took her over to her husband. 'This is Ingrid, from Steinhöring,' she said. Ebner held Clarissa's hand and smiled. He was happy.

It was an appalling experience: this old man on the point of death,* delighted to meet someone over whose birth he had apparently presided, and this so motherly old lady. It was too much. This aged couple were unreal. I thought of what they had been and of what they might have been if . . .

We had less luck with Sollmann's and Ebner's aides.

The men and women personally selected by Himmler to run the Lebensborn organization are naturally terrified of anything that might put them back in the limelight. And like

* Dr Ebner died in the spring of 1974.

Sollmann and Ebner, their subordinates inspire anything but fear today. Frau Inge Viermetz, for instance, who was responsible for the reception of foreign children 'recovered' from Eastern Europe and for placing them with childless SS families, has lost her memory. All she remembers is the charitable activities carried out in the organization's homes. Kurt Heinze, who was in charge of the transport of the kidnapped children, lives in fear. For thirty years this German of Polish origin (he was born at Poznan in 1901) has kept his door shut to all strangers, fearing that one day a victim of the kidnappers might call him to account. It took Gunther Tesch, formerly head of the Lebensborn legal department and now a lawyer at Dortmund, several weeks to decide not to receive us, 'seeing that no good could come of our meeting'.

The SS 'mafia' had got to work. Henceforward no more doors would be opened to us by any leading figure. This was confirmed soon afterwards when we telephoned Max Sollmann and he said: 'Let me give you a piece of advice: drop your project. You won't get anywhere with it. As far as I am concerned, I haven't the slightest desire to reappear in public. I don't want a tile "accidentally" to drop on my head.'

So we turned our attention to the minor figures, those involved in the daily life of the Lebensborns, the staff. We met book-keepers, nurses and supervisors, besides tracking down many women who had had babies in the Lebensborns, as well as children who were born in them or taken to them after being kidnapped. Many of the latter knew nothing about this, naturally enough, and there was, of course, no question of our destroying their lives by telling them things they did not know. The quest took us not only to Germany, but also to France, Belgium, Holland, Poland, Czechoslovakia among other places. In these countries it was much easier, but much more pathetic. In this respect Poland affected us most. We drove thousands of miles through the country to meet parents whose child, or children, had been taken away to be Germanized and never heard of again. Asking them to remember, thus reviving their suffering, was itself a painful experience. And what can be said of the hope suddenly aroused in these people by the unexpected arrival of two foreigners who had travelled so far to see them? They were bound to think we had some sort of information for them, however insignificant. We were so well informed about the organization, we knew so many details about the way in which this child or that had disappeared. How could it be that our knowledge suddenly stopped short at the end of the war in 1945? Alas, generally we knew nothing though, guided as we were throughout those three years by a firm determination not to create new dramas, not to sacrifice the human to the sensational, sometimes we knew something that we could not divulge.

'They only took the most beautiful children.' That was the phrase that remained engraved in our memory after the experience we have just described. Not just because, to a parent, his child is always the most beautiful, but because it implies so many dramas and so much suffering. Then we recalled Frau Ebner's comment on those fair, blue-eyed children. 'Oh, it was the fashion, you know. And in any case they were well treated.'

2. Race and Settlement

> The products of human culture, the achievements in art, science and technology with which we are confronted today, are almost exclusively the creative product of the Aryan. That very fact enables us to draw the not unfounded conclusion that he alone was the founder of higher humanity and was thus the very essence of what we mean by the term man.
>
> ADOLF HITLER, *Mein Kampf*

Having failed to achieve national unity until the second half of the nineteenth century, the élite of the innumerable German kingdoms and principalities had for centuries made the Nordic 'racial' destiny a symbol of German pre-eminence. Thanks to legends, myths and the exploits of ancient heroes, a semblance of German unity could thus be maintained on the basis of the mere principle of pure and superior blood. The frustrated, complex-ridden, petty German states of the nineteenth century, nursing inferiority feelings towards their Western neighbours as these developed into industrial and colonial powers, compensated for their lack of progress and national grandeur by harking back to the past with the aid of the dogma of superior Nordic blood. In view of the lack of a history of Germany which it would have been difficult to write, frustration was handed down from father to son, and until the signature of the Treaty of Versailles in 1871, which gave birth to the First German Reich, this frustration formed part of the national heritage. The only way of combating it was to study the history of races, concentrating on the 'racial' destiny of a people that did not form a single nation.

For a long time this circumstantial philosophy acted as a consolation and a stimulus. But it ended by secreting a popular mysticism that was fraught with danger, being based on a racism that allotted absolute superiority to the Aryan Nordic race descended from the ancient Teutonic kingdoms. Theorists in Germany but also in France and Britain who, in the second half of the nineteenth century, felt a need to devote the greater part of their labours to the Aryan cult and the doctrine of the racial superiority of Germans of Teutonic origin, did the rest. In a country carried along on the wave of nationalist fervour that followed unification, Teutonic legends and pseudo-scientific theories about racial hygiene developed by Ranke, Treitschke, Fritsch, von Luschan, von Ehrenfels, Gobineau, Chamberlain, Ploetz, Schallmayer and many others largely contributed to the development of the future doctrines of the Nazis. By their writings they turned the Aryan myth into a reality only too easily adaptable to the needs of a régime that made racial selection and military discipline overriding principles that legitimized every excess. This process was facilitated by the fact that these men, distorting Darwin's theory of the survival of the fittest, did not hesitate to maintain that the 'struggle of the creative Teutonic-Aryan race' boiled down to the 'struggle against the parasitic Semitic race'.

Thus to all these theorists, philosophers, doctors, psychologists, biologists, and so forth, whether they were scholars, scientists or charlatans, the Teuton necessarily stood for the good, the true and the beautiful, while the bad, the false and the ugly could be the work of inferior, meaning non-German, creatures.

The theory of the master race and of the 'humanity of nature that exterminates the weak for the benefit of the strong', the belief that 'in the old days the victor had the right to exterminate whole races and peoples', were thus Nazi ideas that were accepted by the Germans when Adolf Hitler attained power in 1933. Rauschning states that the resuscitation of the world of Teutonic mythology demonstrated the German capacity to live in two worlds at once by perpetually projecting an imaginary world on to reality. Thus the leaders of the Third Reich, successors to the Nordic heritage spontaneously chosen by their people, used the pretext of purifying the German race to initiate a process of planned reproduction on the one hand and extermination of the other.*

* Hermann Rauschnig, *Gespräche mit Hitler*, Zürich, 1940.

Soon after the reunification of Germany in 1871, Rudolf Virchow, an eminent pathologist and politician, set about compiling national statistics about the colour of the eyes, hair and skin of German schoolchildren. His original idea had been to conduct the inquiry in the army, but the military authorities had refused to co-operate. Virchow was obsessed with the idea of tracing back the Aryan origins of his compatriots, and he believed that the inquiry, undertaken with the enthusiastic aid of the education authorities, would show a large majority of fair-skinned, blue-eyed dolichocephalics, particularly since 'non-Germans and Jews' were excluded.

The inquiry took about ten years, and the head measurements and the colour of the eyes, hair and skin of more than ten million children were duly recorded, together with other information. But all that emerged was that only the inhabitants of north Germany, in other words a minority, could be regarded as being of Nordic stock.

On 6 January 1929 Himmler was appointed Reichsführer SS and put in command of three hundred men who were to be the nucleus of a force on which the Nazi party could rely in all circumstances. In the eyes of its newly appointed commander, that was when the Black Order was born, even before the Nazi accession to power. He regarded the Virchow report as inadequate, first because it was carried out at a time when the technique of racial study was still rudimentary, and, secondly, because it dealt only with children. So far as he was concerned, the racial and demographic requirements of the future Third Reich bore no relation to those of the Reich of 1871. Henceforward the whole nation must be subjected to the principles of selection, and it was necessary to determine as accurately as possible how many German citizens fitted in with the specifications of the authentic Teuton laid down by the official party theorist, Hans Günther: 'Tall, long head, narrow face, well-defined chin, narrow nose with very high root, soft fair (golden-blond) hair, receding light (blue or grey) eyes, pink white skin colour.'* The Reichsführer's view was that heroes of this description would be able to defy all obstacles and carry out the greatest revolution of all.

In accordance with these principles, recruiting offices, supervised by doctors who were experts in 'racial science', set about screening volunteers for genuine descendants of the Indo-European tribes that had emigrated from Jutland (Denmark) and been settled in Germany since the third century B.C. These were to be the stock from which the new Teutonic race was to be bred and the SS was to be recruited. Though the SS was originally founded to provide personal protection for the Führer and assure the security of the party, Himmler envisaged a much more grandiose role. He was aware that in reality, and according to the criteria of the *Rassenprüfer* ('race examiners') of the Rasse und Siedlungshauptamt (Head Office for Race and Settlement, henceforward referred to as the RuSHA), the Germans consisted of very different types – Nordic, Westphalian, Dinaric and even Baltic. But he believed it would be possible to eliminate all traces of impure blood within a few generations, a century at most. He had in mind all those short, dark-haired, dark-eyed Bavarians who had tremendous difficulties in procuring the certificate of Aryan origin, with pedigree dating back to 1750, that had to be produced by every applicant before he could be admitted to the SS. What must Himmler have felt when, in investigating his own family tree, he discovered that his mother's forbears originally came from Hungary and were actually of Mongolian stock? One day he envisaged importing Norwegian girls to Bavaria for the purpose of 'rapidly transforming the Dinaric into a pure Nordic race by means of selective breeding and adequate diet'. In the meantime it was essential to preserve the principle of selection for entry into the SS, confining it to men of use both as combatants and as reproductive males. Thus, in the early months of 1933, very strict standards indeed were being applied to the selection of the 50000 men accepted by the 'racial experts'. Only individuals in whom perfection of movement went hand in hand with perfect physique could

* Hans Günther, *Ritter, Tod und Teufel*, Munich, 1928.

aspire to transfer from the German to the Germanic race. The first examiners were so severe that the Reichsführer was able to state that 'until 1936 . . . we accepted no one who had even a filled tooth. The men we assembled in that early Waffen SS were the most magnificent specimens of manhood.'

Volunteers were deterred neither by the rigorous terms of enlistment – four years for men, twelve for NCOs, twenty-five for officers – nor by the strict requirements in regard to personal appearance, race, antecedents, 'outlook on life', and so on. Unemployed men and peasants' sons, attracted by the prospects of a career open to anyone with 'talent' in an organization with no class distinctions, rapidly swelled the ranks of this *corps d'élite*; too rapidly in the eyes of the Reichsführer, who suspended recruiting for nearly two years from 10 July 1933. Sixty thousand 'problem' cases were thus disposed of: alcoholics, homosexuals, professional unemployed and, above all, Aryans of doubtful origin. The Reichsführer himself, or, failing him, the chiefs of the race offices, inspected photographs of every applicant to make sure his face bore no sign of taint, such as 'orientally' prominent cheek-bones, 'Mongolian' slit eyes, dark, curly black hair, legs too short in relation to the body, a body too long in relation to the arms, a bespectacled 'Jewish' intellectual look. The list of defects that formed an insuperable barrier to becoming a paladin of the Führer was long and carefully compiled. Himmler had decided that man was no longer to be descended from the ape, but from the SS. He was to have one leader, the Führer, one country, the Reich, and one religion, blood. He was to be tall, fair, blue-eyed, athletic, and worthy of the SS motto, 'My honour is loyalty.'

The question put to him before he took the oath was: 'Why do you believe in Germany and the Führer?'

To this the reply was: 'Because we believe in God, we believe in the Germany that He created in His world and in the Führer Adolf Hitler whom He sent us.'

The oath itself was as follows: 'I swear loyalty and courage to you, Adolf Hitler, as Führer and Chancellor of the German Reich. I vow obedience to the death to you and my superiors appointed by you. So help me God.'

The principle of racial segregation propounded by Hitler and numerous German thinkers before him left the domain of theory and entered that of practice with the introduction of the SS pedigree book (*Sippenbuch*). This led to a flurry of activity in clergymen's offices and civilian register offices, faced with an influx of men wanting to establish the details of their family trees. Candidates for the Black Order had to prove unsullied Aryan ancestry dating back to the Thirty Years War. Those who passed the test were doubly blessed: they got a job with, above all, a 'uniform that did credit to the German nation'; and at the same time they began to fill the ordinary German citizen with alarm. This did not trouble Himmler. 'I know there are people in Germany who feel ill when they see our black tunic,' he said in 1936. 'We understand this, and do not expect to be liked by too many people.'

The SS, he declared, was 'a body of German men of the Nordic type selected in accordance with special principles', and anything damaging to its prestige annoyed him. But in spite of the innumerable proclamations by the top leadership, in spite of all the propaganda films illustrating the superiority of the new Germanic hero and the screen dramas in which modern Siegfrieds invariably ended by discovering traces of their Aryan ancestry, various Ministries of the Reich tried to tone down Himmler's Nordic ardour. In view of the predominance of the dark over the fair, and of the short over the tall, among the German population at large, memoranda were from time to time circulated asking for less public emphasis to be laid on Nordic superiority in relation to the rest of the population.

Himmler took no notice; there was no abatement of the flood of propaganda about race and pedigree emanating from the RuSHA.

This organization was established on 31 December 1931, and one of its first tasks was to draw up an official list of 'racial values'. Originally this was used only for sifting recruits for the SS, but after the outbreak of war it was extended to dividing human beings into two categories: those worthy or unworthy of belonging to the German Reich.

The RuSHA racial propaganda machine moved rapidly into high gear. It put out slogans and pamphlets, published a periodical, the *SS Leithefte*, and arranged lectures at which novices – on whose shoulders responsibility for racial selection throughout Europe subsequently rested – were indoctrinated with the idea of Nordic superiority, which was inevitably linked in the minds of the 'race experts' with the idea of a return to 'Teutonic usages and customs'.

Soon slogans started appearing on the walls of towns and villages, warning the population against the dangers involved in mixing true Aryan blood with that of inferior races. The Nazi leaders soon had their reward. No public protest followed the publication in 1935 of the Nuremburg laws, which were intended 'to preserve German blood from all contamination and protect the German race to the end of time'. These forbade marriage and sexual relations between Jews and citizens of German blood and, pending a 'Final Solution', deprived Jews of all civil rights.

According to the Nazi leaders, the progressive deterioration of the German race, the inability of twentieth-century man to produce pure-blooded and racially valuable children, was due to the fact that inferior races, the Jews in particular, had been permitted to mix with the superior races. Every effort was therefore to be made to reverse this 'degeneration of the species', which was discernible right up to the highest levels of the State. By Himmler's orders, the RuSHA laid down a series of principles: insistence on racial hygiene, improving the racial stock by means of selection, supervision of the marriages of individuals of pure blood, and the bringing up of children in State institutions.

Many German theorists before the Nazis had proposed measures to 'purify' the German race, but the race experts of the RuSHA relied most heavily on the works of the Bavarian Dr Wilhelm Schallmayer, who had won a prize in 1900, offered by the Krupp Foundation in connection with celebrations for the beginning of the twentieth century, for the best book on the superiority of the German race. One of the subjects with which Dr Schallmayer dealt was 'Racial hygiene and its control in the national interest', and the title of his book was *Inheritance and Selection in the Life of Nations*. It made eight main points:

1. that young people should be made to realize that there is no nobler career for a girl than that of wife and mother;
2. that a woman's social position should depend on the number of children she has borne;
3. that a cult of the family should be created and developed;
4. that men should be encouraged to marry young;
5. that the inheritance laws should be reformed to conform with the population policy;
6. that work by women outside the home must be reduced to a minimum;
7. that sexual disease must be fought; and
8. that special homes should be created for home-coming soldiers, by which preference was to be given to those against whom there were no objections on the grounds of racial hygiene and from whom an improvement of the birth rate was to be expected.

The Krupp prize-winner, drawing on the work of other theorists on the subject of reproduction, also suggested that 'returning warriors (who are generally characterized by above-average racial qualities) should be allowed more than one wife'. Quantitative considerations did not cause him to neglect the qualitative aspect; in 1900 he was already proposing 'permanent eugenic control of the German population'; 'family books' would be compiled from 'personal' genetic records for every individual concerned. The doctors responsible for the racial and biological examination of individuals would be subject to professional secrecy, except towards the State, and the 'family books' would make it possible to supervise and, if necessary, forbid prospective marriages.

This Bavarian doctor, a true precursor of the Lebensborns and the concentration camps, insisted that an essential part of any breeding policy was to ensure that the most unsuitable heritable variations were not reproduced. With this in mind, he envisaged not only forbidding certain marriages but also introducing sterilization, because 'sex life is not just a private affair, but should be a sacred matter dedicated to a higher cause'.

In a circular issued by the RuSHA the terminology was modernized:

The value of the genealogical tree now resides no longer merely in its showing one's Aryan descent; it offers us guiding lines for the breeding of a new race of leaders. This is of the greatest importance for the SS in particular if it is to maintain its character as an élite and an Order.

So much for the selection of males. As for the selection of females, the *SS Leithefte* stated:

Every SS man should seek out for himself a wife of at least equal or if possible actually higher racial value . . .

A passage devoted to the SS in a biology textbook for girls' secondary schools (published in 1943) says:

Ever since the time of struggle the SS as a leading organization of the party has made it the SS man's duty to choose a biologically flawless wife and, moreover, has constantly reminded him of his obligation as a member of an élite formation to have a large number of children.

How was this obligation to be fulfilled? For the time being Himmler saw only one way. On 1 January 1932 an 'engagement and marriage order' was issued to the SS, and was later extended to the whole police force by Himmler's decree of 27 July 1938. It included the following:

Every SS man intending to marry must obtain the approval of the Reichsführer SS.

A member of the SS who marries though approval has been refused will be struck off the roll. He will be given the opportunity to resign.

Marriage applications will be dealt with by the SS Race Office.

The SS Race Office will keep the SS pedigree book in which the families of members of the SS will be recorded after a marriage has been approved or consent to an application for an entry has been granted.

The Reichsführer SS, the head of the Race Office and its staff are honour-bound to secrecy.

Approval of a marriage was generally granted only after the submission of a questionnaire that had to be carefully filled in by the SS man and his fiancée. The RuSHA race experts added their comments, based on about twenty physical characteristics, including the applicant's height, standing and seated; the shape of the skull, face and forehead; colour and location of the eyes and distance between them; length, breadth and curvature of the nose; length of arms, legs and body; colour, growth and quality of body hair; skin colour; back of the head, cheek-bones, lips, chin, eyelids; thorax (male applicants), pelvis (female applicants). In addition to all this, the SS man's bride had to be able to provide evidence that neither she nor her parents suffered from any physical or mental disease. Also she had to submit to an examination by SS doctors to make sure she was not sterile. Finally, provided she overcame all these hurdles, she had to produce a family tree showing there had been no Slavonic, let alone Jewish, blood in her family since 1750.

In a note in the files directed to his 'race expert' doctors, Himmler wrote:

> The Reichsführer will in future review questionnaires about mothers under three heads:
> 1. Those in complete accordance with SS selection principles.
> 2. Those of good average standard.
> 3. Those not suitable for selection.

A new wave of photographs inundated the Reichsführer's desk. Males, naked or in trunks, who had applied to the SS recruiting offices, were followed by young women applicants for marriage permits. They had to submit full-length photographs to the RuSHA, in bathing costume if possible. Their fiancés having done the same, the 'race experts' subjected the question of their future harmony to microscopic examination. The final decision rested with the Reichsführer.

To avoid mistakes which might subsequently occur in the selection of subjects suitable for 'Germanization', the RuSHA in 1942 distributed a pamphlet, *The Sub-Human*, to those responsible for that selection. 3 860 995 copies were printed in German alone and it was translated into Greek, French, Dutch, Danish, Bulgarian, Hungarian and Czech and seven other languages. It stated:

The sub-human, that biologically seemingly complete similar creation of nature with hands, feet and a kind of brain, with eyes and a mouth, is nevertheless a completely different, dreadful creature. He is only a rough copy of a human being, with human-like facial traits but nonetheless morally and mentally lower than any animal. Within this creature there is a fearful chaos of wild, uninhibited passions, nameless destructiveness, the most primitive desires, the nakedest vulgarity. Sub-human, otherwise nothing. For all that bear a human face are not equal. Woe to him who forgets it.

The vistas opened by the establishment of a Reich that was to last a thousand years involved the RuSHA in a specific task. The former chicken-breeder Heinrich Himmler was given the opportunity to make an old German dream come true – the dream of purifying and renewing the Germanic race.

3. Bearing Children for the Führer

'The human garden consisted of a country settlement with about a thousand women and a hundred men chosen by specialists for their racial superiority to the average civil community. . . . Three hundred of these colonies would produce a yearly influx of a hundred thousand unbroken human beings and cause a turn of the tide throughout the life of the nation.'

DR W HENTSCHEL, *Mitgart. Ein Weg zur Erneuerung der germanische Rasse*

The First World War had decimated the ranks of the young men in all the principal belligerent countries. In Germany the disproportion between the numbers of young men and young women was perhaps even greater than elsewhere. Only one of every four women between the ages of twenty-five and thirty could hope to find a husband.

The consequence was a big drop in the birth rate, aggravated in the post-war period by unemployment, inflation and fear of the future. The spread of birth control and the emancipation of women, which had made big strides in Germany in the early years of the twentieth century, also conflicted with the demographic needs of a country bled white by four years of war. The alarming aspect of the population statistics published a few years after the war provided the Nazi leaders with a psychological atmosphere that enabled them to impose a policy which no one could refuse. According to a Nazi Party report published in 1936, births in Germany between 1915 and 1933 showed a drop of 14 million in comparison with the period between 1896 and 1914. Still more serious was the fact that the German population of 37 million in 1885 had produced the same number of children as the population of 67 million in 1935.

The result was a complete change in women's role. Every effort was made to induce them to return to the old-fashioned domestic German virtues in which woman was the guardian of the hearth. The most important part of a woman's life suddenly became that in which she could bear children. If she was fertile, pregnant by order of the Führer, everything was due to her. If she was intellectual, independent, childless, she was exposed to the violent onslaughts of the breeding experts.

The campaign against childlessness in which the Nazis engaged immediately after their accession to power was first concentrated on the emancipated woman. She was to be eliminated at all costs from public life, forced back into the home in accordance with the old principle of the three Ks – *Küche, Kinder, Kirche* ('kitchen, children, church'). At a time when German women had been given the vote and were struggling for liberation from male domination, the Nazi Party published in January 1921 a statement in which they undertook to exclude women permanently from all important positions in politics. Hitler in particular spoke of the emancipation of women as an 'unnatural symptom, like parliamentary democracy'. According to the party leaders, such tendencies were evidence of frustration due to inadequate functioning of the sex glands. Goebbels compared women to animals:

> A woman's duty is to be attractive and bear children. The idea is not as vulgar and old-fashioned as it might seem. A female bird makes itself beautiful for its mate and hatches out her eggs for him. The removal of women from public life . . . is only to restore their female dignity.

As early as 1933 changes were made in school curricula with a view to preparing girls for home life instead of university. Yearly, almost monthly, new decrees were issued diminishing women's status. According to the Nazis, women were only able to make decisions emotionally and were thus unqualified to hand down justice, among other things. In June 1936 they were accordingly banned from judicial office, as well as from certain other posts in the judicial administration, including that of public prosecutor. Married women doctors were deprived of their jobs 'in view of the necessity of devoting themselves to childbearing', but this was revoked when the shortage of male doctors made it necessary to re-employ them.

In spite of all this discrimination and the attendant disadvantages, most women, surprising though it may seem, considered the maternity cult a good thing, a necessity in the national interest. To demonstrate the genuineness of their pro-Nazi, anti-bourgeois feelings, they did their best to look as much as possible like the Nazi ideal woman: a striking blonde with big hips and hair tied back at the nape of the neck or plaited on top of her head. Make-up was 'un-German' and only for faces 'marked by the eroticism of eastern females'. In Berlin, women wearing make-up were insulted on buses; they were called whores, and sometimes even traitors.

Slimness was also condemned. It was taken for granted that women who were too slim could not have many children, just as those excessively devoted to their dogs were robbing their future children of the love to which they were entitled. These points were continually hammered home in films. Pictures of domestic pets were invariably accompanied by comments such as: 'Those who give a dog the place to which a child is entitled commit a crime against nature and against our people'; or, 'Excessive love of animals is contrary to nature. It does not elevate the animal, it debases the human being.'

Propaganda directed at women, though, of course, it varied according to circumstances, was always inspired by two distinct aims: numerical quantity and racial quality. The former led directly to the campaign on the 'birth front', to meet the growing population needs of the country; the latter was restricted to the élite of the German nation, the blond blue-eyed Nordics, and led to what was later known as the SS human stud-farms.

So far as numerical productivity was concerned, from 1933 onwards large families of German stock were entitled to innumerable benefits. Hitler's mother's birthday – 12 August – became the Day of the German Mother, when public ceremonies took place at which mothers of large families were decorated with the German Mother's Cross – bronze for four to six children, silver for six to eight, and gold for eight or more. Women in possession of this award were entitled to the public respect enjoyed by front-line soldiers. The official view in the matter was that the dangers to which a mother's health and life were exposed in thus serving her people and her country were equivalent to those to which the soldier was exposed in battle.

By August 1939 three million German mothers had been decorated and belonged to what the man in the street called the 'Order of the Rabbit' (*Kaninchenorden*). When wearing their decoration these women were entitled to the Hitler salute from members of the party youth organizations, as well as to all sorts of special allocations, allowances and other privileges. During the war these often took the unexpected form of entitlement to a housemaid chosen from the millions of girls and women from Eastern Europe who were sent to forced labour in Germany. Polish, Russian, Czech and other female deportees were distributed among deserving families 'so that the fertility of the German mother should not be diminished by physical work'. So great was the preoccupation with fertility that of the 18 million German mothers who were expected to devote themselves exclusively to the bringing up of their children, not one was sent to a factory before 1943. The party leaders never tired of reiterating that war was less harmful to a nation than a decline in the fertility of its women.

Women having been restored to the ideal condition of wife and mother, unmarried mother or fiancée, a pitiless struggle began against homosexuality, prostitution and abortion. Publicity for contraceptives was banned from the start. Interrupting the pregnancy of an Aryan German woman was considered an act of sabotage of Germany's racial future and was punished accordingly. Under the Weimar Republic the maximum penalty for abortion had been a fine of forty marks. The Nazi courts sentenced doctors guilty of the offence to from six to fifteen years' imprisonment, generally served in a concentration camp. Later the sentences were even more severe, but 'persuasion' bore fruit. The abortion rate, from being nearly a million a year before 1933, fell to less than 200000 in 1938. Abortions were officially classified as miscarriages.

The campaign against homosexuality also began in 1933. Like everything else, it was undertaken on a grand scale. Propaganda films were distributed at discreet intervals, youth leaders were supplied with prepared speeches on the subject, decrees were issued, and repressive measures were taken by the party, whose moralists were certainly less disturbed by homosexual activity than by the childlessness that ensued.

The campaign was most vigorously pursued in the SS; it reached its height with the publication by Himmler on 15 November 1941 of a decree laying down the death penalty for any member of the SS or the police who engaged in sexual relations with another male. Most of those found guilty, if they escaped the death penalty, ended in concentration camps. Indeed, Himmler had his own nephew, SS Obersturmführer Hans Himmler, 'liquidated' at Dachau for homosexuality.

For similar reasons, because of their failure to be involved in the national effort to increase the population, childless couples, sterile women and even women 'too old for child-bearing' were made a target of the Nazi breeding experts. Of 16 million married women at the end of 1938, 22 per cent were childless; in Berlin the percentage was 34·5. This was in spite of the fact that, up to the beginning of the war, the so-called marriage loan was one of the most attractive items in the régime's population policy. Each couple was lent 1000 marks, a quarter of which was converted into a gift from the State on the birth of a child. Thus the birth of the fourth child wiped out the loan. By the end of 1938 the Reich Finance Ministry had made 1212000 marriage loans, of which nearly a million no longer had to be repaid.

Sterile marriages led to the 1938 divorce law under which, among other things, a wife's sterility or refusal to have children became valid reasons for divorce. The results were not long in coming. Divorces leapt from 42000 in 1932 to 62000 in 1938. Less than two years after the law was promulgated, 80 per cent of the husbands who repudiated their wives did so because of the latters' 'inability to give them children'. Three out of five of the wives concerned were over forty-five and had been married for more than twenty years, and a husband was able to declare his wife 'incapable of childbearing' even if she had borne several children when young. A mere declaration by the husband that his wife was sterile was sufficient to enable the courts to grant a divorce without further formality.

Before the start of the war the SS had come to see the Christian idea of marriage as a principal enemy of fertility. The war itself confirmed them in their view. As Dr Felix Kersten, Himmler's masseur and the recipient of his confidences, recollected one of their conversations:

Increasing losses on the Eastern Front were causing Hitler great concern, so Himmler told me, since they were consuming the substance of the German people and must be replaced if the German people were to keep the conquered territory. The Führer had therefore decided to make a fundamental change in the existing law of marriage immediately after the war and introduce bigamy.

My personal view, Himmler added, is that our development is leading to a breach with monogamy. The present form of marriage is the satanic work of the Catholic Church, the marriage laws are themselves immoral ... In a bigamous marriage each wife would act as a spur to the other to approach the ideal in every respect ... As it is normally impossible for a man to manage for a whole lifetime with one woman only, infidelity, and, to conceal it, hypocrisy, is forced upon him. The consequences are married strife and mutual antipathy, and the final outcome is lack of children. In this way millions of children urgently needed by the State remain unborn.*

Himmler himself had left his wife Marga to live openly with his mistress Hedwig, by whom he had two daughters. Martin Bormann, Hitler's deputy, was in a similar position, Frau Bormann being so dedicated to the lofty aim of childbearing that she had congratulated her husband on winning the favours of a young actress. The one thing she asked of

* Felix Kersten, *Totenkopf und Treue*, Hamburg, 1953.

him was that he should take care that they were never pregnant simultaneously, so that he would always have one of them available.

Following the outbreak of war a further wave of statements, pamphlets and orders had descended on the German wife and mother. On Christmas Day 1939, Rudolf Hess had published in the *Völkischer Beobachter* an 'Open Letter to an Unmarried Mother', stating the new morality as follows:

> The Nazi philosophy of life having given the family the role in the State to which it is entitled, in times of special national emergency special measures can be made different from the basic principles. In wartime, which involves the death of many of our best men, every new life is of special importance to the nation. Hence if racially unobjectionable young men going on active service leave behind children who pass on their blood to future generations through a girl of the right age and similar healthy heredity . . . steps will be taken to preserve this valuable national wealth.

The climax to several years of kicking around traditional morality had already been reached by 28 October 1939 when Himmler published his order to the SS and the police:

> Every war involves a tremendous letting of the best blood Many victories won by the force of arms have been a shattering defeat for a nation's vitality and blood. But the sadly necessary death of the best men, deplorable though it is, is not the worst. Far worse is the absence of the children who were not born of the living during the war and of the dead after it.
>
> Beyond the boundaries of bourgeois laws and customs that are perhaps normally necessary, it will be a high duty of German women and girls of good blood to become the mothers, outside the boundaries of marriage, and not irresponsibly but in a spirit of deep moral seriousness, of children of soldiers going on active service of whom fate alone knows whether they will come back again or die for Germany. . .

The Lebensborn correspondence as a whole includes an impressive number of letters signed by Himmler or Brandt, his aide-de-camp, on women's fertility, the aptitude for childbearing of one woman or another. Himmler once sent the following 'note' on 'the begetting of a son' to Max Sollmann, the head of the Lebensborns:

When the subject of begetting boys or girls cropped up in conversation with SS Obergruppenführer Berger recently, he told me that in his Swabian homeland when a family definitely wants a boy at last, the following is the custom. Husband and wife refrain from alcohol for a week. Then the husband leaves home at midday and walks to Ulm twenty kilometres away and walks back again. He must not stop at any inn on the way. For a whole week before the day of conception the woman does no work, eats very well, sleeps a great deal, and does nothing whatever to tire herself in any way. The begetting takes place after the husband's return from his walk. The result is said invariably to be the birth of a boy.

In another letter Brandt quoted the case of a woman of forty-nine or fifty who was expecting her third child. The first child had been born three or four years previously and she had had a miscarriage many years before. According to the information given by her, she had not had her first period until the age of nineteen. Himmler asked for an investigation into this case of retarded maternity.*

Himmler advised an SS officer's wife expecting her husband home on leave to 'have yourself examined by a gynaecologist with a view to determining the most favourable time for conception, for only a leave taken at the right time can give you some assurance, according to human predictability, that you will achieve your and your husband's desire to have a child'. In fact Himmler took a consistent interest in the question of arranging leave at the right time. He invited many SS officers' and men's wives to spend five or six days at the public expense near where their husbands were stationed so that their marriages might result in the children that were so desirable and necessary. He personally also drew up a list of suitable hotels.

In a memorandum dated 29 January 1944, Martin Bor-

*Letter from Obersturmbannführer Brandt to Dr Ebner, chief medical officer of the Lebensborns, 4 August 1942. The woman referred may have been the wife of the concentration camp doctor Sigmund Rascher (see page 113).

mann estimated that once the war was over Germany would have a surplus of from three to four million unmarried women with no hope of finding a husband. 'Every German woman', he went on, 'must be able to give birth to as many children as she wants if in twenty years' time the Reich is not to be short of the divisions which will be a vital necessity to us if our nation is to survive.' The memorandum called on German males 'to maintain stable conjugal relations with one or more women', and, among other things, recommended the working out of a complete and detailed propaganda campaign to persuade women of the wisdom and prudence of this idea.

Obviously those most affected by this policy of teeming population growth were the young. On 28 October 1935 the leaders of the youth movement had announced the institution of 'biological marriages' to give respectability to extramarital sexual relations between young people who shared the same ideals. The propaganda was chiefly directed at girls, for the simple reason that there were so many surplus women in Germany. Hence the leaders of the Bund Deutscher Mädchen, or League of German Girls (BDM), continually reminded girls at mixed meetings that, though they would not all be able to find husbands, all would be able to become mothers.

The putting into practice of this new principle in the service of the Führer worked havoc from the outset, and one of the consequences was that thousands of copies of an 'Open Letter to Goebbels', signed Michael Germanicus, circulated clandestinely throughout the country; many people were sent to prison for having it in their possession. The pseudonymous author said, in effect, that an incalculable number of sexual crimes were committed at youth camps, and that at meetings of the Hitler Youth and BDM girls of sixteen and even fourteen had been morally and physically ruined. No statistics are available, but it is known that during the 1936 celebrations at Nuremberg of the anniversary of the foundation of the Nazi Party, attended by more than a hundred thousand young people, nearly a thousand young girls became pregnant. In the early years of the régime, the indignation of many parents made them try to trace the youths who had fathered their daughters' children during the celebrations, but, in spite of the admirable organization of the Hitler Youth, only half of them were found.

Nazi character training and the sexual education that went with it thus largely took place at the camps, outings and sporting and quasi-military meetings to which the régime attached so much importance. Children no longer belonged to their parents, but to the Nazi movement. In the event of conflict, such as, for instance, might arise from the pregnancy of a girl of fourteen, the Hitler Youth always had the last word. The girl had only been a good pupil and taken in what she had been taught at school. A typical dictation entitled 'Biological Marriage' was as follows: 'We shall all today or tomorrow be able to abandon ourselves to the rich emotional experience of procreating in the company of a healthy young man, without troubling about the impediments that encumber the antiquated institution of marriage.' She had learnt the lesson inculcated into her.

In spite of the opposition of families with religious and moral traditions, the propaganda had swiftly borne fruit. Girls justified their behaviour by phrases such as: 'But the Führer wants us to have children; or: 'I'm a German mother; what do you want of me?' or: 'The others do it too.' From puberty onwards German girls and boys regarded sexual relations as a form of sport that posed no problems. From 1937 onwards, the number of illegitimate births substantially increased. By order of the Ministry of the Interior, unmarried mothers were entitled to be addressed as 'Mrs' and they and their children were given all the rights of legitimate mothers.

An idea of the consequences of the campaign 'for liberating sexual relations with a view to repopulating Germany' can be obtained by glancing at the classified advertisement columns of the newspapers before and during the war. Eight out of ten announcements in the marriage columns were of the type: 'Girl with child seeks comrade for life with view to marriage.' A frequent death announcement, signed 'The fiancé, the son and the family', was: 'Our fiancé and dear father died on . . . for the Führer and Greater Germany. He will continue to live through his son.'

To facilitate marriage when the girl had fulfilled the Führer's desire by becoming pregnant, the authorities provided official forms for the use of 'persons of the male sex desirous of obtaining a reduction in the age of majority'. Such young men were allowed to marry without their parents' consent.

In fact, in view of the progressive destruction of the family unit, the obtaining of parental consent had long been no more than a formality.

Boys called up by the Hitler Youth, the Labour Service or the armed forces were continually away from home. Girls between fourteen and seventeen were similarly kept busy by the BDM, while between seventeen and twenty-one their time was taken up by the Faith and Beauty organization, Mothers' and Children's Aid, or as army auxiliaries. Thus, off duty, they were left to their own devices. The crime rate among the young quickly rose alarmingly. Of about 500000 sentences imposed by the courts in 1933, 16000 were on persons under the age of eighteen. In 1939 the figures were 21000 out of 300000.

Surprising though it may seem today, the campaign to encourage the unmarried to have children led to the only street demonstration to take place in Germany during the twelve years of the Nazi régime. This occurred on 19 February 1943, after the Bavarian Gauleiter Paul Giesler had made a speech to the students of Munich University. After accusing the girls of attending lectures 'for the sole purpose of finding a husband', he went on to enlarge on the Nazi idea of sex for the purpose of propagating the German species. He reminded the students of the duty of every German woman to bear children for the Führer, and then invited them to play their part in the repopulation of Germany: 'If you have no one with whom to conceive children. I'll lend you my men . . . I promise you you won't be disappointed.'

Shocked at this crudeness, students of both sexes went out into the streets, shouting anti-Nazi slogans. They dashed through the Schwabing quarter, scrawled 'Down with Hitler' on the walls, and handed out leaflets to the hostile crowd, calling on them to rebel against the dictatorship. Some of the demonstrators belonged to a clandestine organization called the White Rose, which for several months had been engaged in anti-Nazi activities in various German universities. Their leaders were a medical student named Hans Scholl and his sister Sophie. The two were recognized by a *Blockleiter*, who denounced them to the Gestapo. After three days of interrogation and torture, they were sentenced to death by the People's Court and were beheaded, with several of their comrades, one of them a professor, on 22 February 1943.

Despite all the propaganda and offers of awards and material advantages, the *legitimate* birth rate had increased only slowly in the first few years of the régime. The Nazi race defenders then made an important discovery, the best results were to be obtained from unmarried girls of Nordic type, who were much quicker to adapt to current realities than married women, even including the wives of members of the SS. To these girls childbearing did not just involve the right to dispose freely of their persons; they felt it their duty to play their part in the multiplication of the race. To them biological marriage in the service of the Führer became a State institution in which the male was only a reproducer, while woman was a living source of pure blood.

The race experts considered it necessary to exercise selection on very definite lines. Racial requirements and control of their application necessitated an 'aseptic' environment to prevent the future super-race from being tainted with impure blood. The administrative task involved was entrusted to the SS, and it became the specific responsibility of the race and settlement offices founded by Heinrich

Himmler. By the mid 1930s centres were being set up to enable racial selection and human stock-breeding to go hand in hand with the political indoctrination of future generations. These 'sources of new life', controlled and supervised by the Nazi élite, were the Lebensborns.

4. The Beginnings

'The eternal law of nature to keep the race pure is the legacy that the National Socialist movement has bestowed upon the German people for all time. In this spirit the community of the German people marches into the future.'
Propaganda film, Berlin, 1935

At the instigation of the RuSHA, and in agreement with various organizations for the protection of mother and child, the Lebensborn Registered Society came into existence on 12 December 1935. It provided a crèche for children and facilities for the confinement of wives, fiancées and girl friends of SS men and members of the police. The new organization at first led a rather precarious existence, for there had always been a number of maternity homes for unmarried mothers in Germany, and new ones were established after the *Führerdienst* came into official existence.

The Lebensborn organization was established by Himmler's orders to ensure that after 'the victory of the German spirit, the victory of the German child' would follow. In other words, it was to meet the 'urgent need' to enable 'racially valuable' unmarried mothers to have their babies without their parents' knowledge and, if they so desired, hand them over to the SS, who would care for them and arrange for their adoption. Pregnant girls, conscious or unconscious victims of Nazi propaganda, began to take advantage of this. As a result of the heavy penalties for abortion, many upper-class women also availed themselves of the opportunity of thus ridding themselves of unwanted illegitimate children, and according to Paula Hessler, the former Lebensborn nurse, their number was very high. At all events, the authorities were delighted with the project's success, and did their best to stamp out the last remnants of the atavistic fear of the pastor or priest still felt by unwed mothers in Germany, as elsewhere. As an issue of the notorious *SS Leithefte* pointed out:

> Having many children is the real meaning of life. That is why a man who has become the father of *healthy* children is worth more to us than all the deliberately childless priests of a whole millennium. And a German woman who has given life to many children and has looked after them for a lifetime is to us a phenomenon more sacred than all the childless nuns our nation has ever produced.

Such high esteem was reserved only for 'racially valuable' mothers, and the word 'healthy', of course, meant 'racially' healthy. However, Himmler was concerned with qualitative rather than quantitative success, and he found himself faced with a threat that had to be promptly averted. All too often his men chose to ignore the constantly reiterated warnings against the dangers to them and their offspring involved in the choice of a non-Nordic woman, and the quality of the future mothers, wives, fiancées and girl friends chosen by them left much to be desired. Himmler frequently complained to his entourage that Nordic girls were neglected by SS men because they were more frigid and less frivolous than girls of inferior racial quality. As he said when addressing a Hitler Youth meeting on 22 May 1936: '... You can see it even more simply in the dance hall – the racially less valuable girl was asked to dance because she was, let us say, more attractive, and the racially more valuable girl was a wallflower the whole evening ...'

There were even SS men, as an officer in the *SS Leithefte* noted, who 'still marry short, squat girls with round figures'. According to the writer of the article, unions of this kind were undesirable for the following reasons:

1. The children would be of unpleasing [*unharmonisch*] appearance.
2. There could be difficulties at birth, for there was often a disproportion between the child's size and the birth passage.
3. Glandular, hormonal, disturbances are often present in such small women, particularly if they are also greatly overweight. In such cases the ability to conceive is greatly hampered, or the women are actually infertile. Also women of marriageable

age who are greatly overweight are usually unattractive and in no way correspond to our Nordic idea of beauty and thus the SS ideas of selection.

For such reasons, more than 50 per cent of the pregnant girls or women who applied for admission to the Lebensborns were turned away.

Himmler was to remain puzzled all his life by the problem of why men fell for the non-Nordic, more sexy, type of girl, and right up to the collapse of Nazi Germany the problem of recruiting the right type of future mother was one of his major preoccupations.

On 3 July 1944 he issued the following order:

1. I want to see no more applications to marry women who do not comply in every respect with the racial criteria in force.
2. Every problematical marriage application must be submitted to the Reichsführer, who alone has authority to reject it.
3. Every marriage application involving a non-German but Teutonic citizen must be submitted for approval to the Reichsführer.

From the beginning Himmler, who assured his entourage every morning that he was 'delighted every time a child is born, never mind how it was conceived', had taken personal charge of the destinies of the Lebensborn organization, certainly intending to be aided in his task by the members of the SS as a whole. In a circular dated 13 September 1936 he said:

The SS has taken a first step in issuing the marriage order. But a marriage without many children is no more than an affair. I hope here too members of the SS, and particularly its leaders, will set a good example. Four children are the necessary minimum for a good and healthy marriage. In the event of childlessness it is the duty of every SS leader to adopt racially valuable children free of hereditary illnesses and to bring them up in the spirit of our ideas and give them an education appropriate to their abilities. The Lebensborn organization will aid SS leaders in the selection and adoption of racially valuable children. The Lebensborn Society is under the direct personal control of the Reichsführer SS, it is an integral part of the Race and Settlement Head Office, and its objects are:

1. To support racially and genetically valuable large families.
2. To accommodate and look after racially and genetically valuable expectant mothers who, after careful investigation of their families and those of the fathers of their children by the Race and Settlement Head Office, can be expected to give birth to equally valuable children.
3. To look after these children.
4. To look after the mothers of these children...

Less than a year after the establishment of the first home by the RuSHA, Himmler's order of 13 September 1936 put the Lebensborn Society directly under the SS general staff. Henceforward it would be known to the SS administration as the 'L office', and it came under the Reichsführer's direct supervision.

Himmler had thought of everything, as is shown by the first prospectus vaunting the merits of the Lebensborn organization:

The expenses involved in carrying out these tasks will be met in the first place by members' subscriptions. Every SS leader attached to head office is honour-bound to become a member. Subscriptions are graded according to the SS leader's age, income and number of children, and it is taken for granted in this connection that he will have married by the age of twenty-six. If he is still childless at twenty-eight, a higher subscription will be due. At the age of thirty his second child should have arrived; otherwise his subscription will again be increased. If at the appropriate ages further children have failed to appear, corresponding increases in the subscription will again become payable. Those who believe they can escape their obligations to the nation and the race by remaining single will pay subscriptions at a level that will cause them to prefer marriage to bachelordom.

All SS leaders not attached to head office and all SS men will – to the extent that they are able – join the society to help realize in a comradely manner this great task of the SS racial stock.

The Lebensborn ... intervenes helpfully wherever the need arises of preserving and as far as possible promoting life valuable to our people. For, if the State now precludes from reproduction that section of the population that is afflicted with hereditary illnesses, as a counterpart to that all genetically healthy life of good blood that comes into the world must at all costs be aided and preserved.

There was another reference in the prospectus to the ban on reproduction imposed on inferior forms of humanity, and it again insisted on the duty of the master race to reproduce itself and multiply.

One of the things that Himmler liked to say was:

> Supposing Bach's mother, after her fifth or sixth or even twelfth child, had said that'll do, enough is enough, the works of Bach would never have been written. The same applies to Richard Wagner. He was a sixth child. I tell you our culture would have been tremendously impoverished if in the old days families had generally been limited to four or five children.

The first Lebensborns, now under the direct supervision of the SS Führungsamt, were reorganized in 1936. A medical superintendent was put in charge of each home, aided by a head nurse, an administrator, and a secretary, all members of the SS or of the party. SS Sturmbannführer Pflaum (succeeded in 1940 by Max Sollmann) was put in charge of the Head Office. In each home the medical officer was in charge, subject only to Pflaum and Himmler. He acted as registrar of births and deaths and was responsible for keeping order in the establishment and for victualling it. Also he was in charge of the files, which were kept under lock and key and to which he alone had access. The stamp used on all correspondence was that of the Reichsführer's personal staff.

Henceforward these establishments functioned according to the *Führerprinzip*, the leadership principle to which the whole country was now subject. The first homes at the entrance to which the SS black flag fluttered were at Steinhöring, near Munich, Wernigerode in the Harz, and Klosterheide, near Lindau, in the Mark. Bad Polzin, in Pomerania, was opened a few months later. The prospectus already quoted mentioned that, in view of the importance of the matter to the national population policy, the construction and installation costs were to be met by the NS-Volkswohlfahrt, the Nazi welfare organization.

Himmler knew that most of the children born in these homes would be illegitimate. Hence the emphasis from the outset on the necessity of maintaining the most absolute secrecy.

> We call on doctors to maintain secrecy, not only for professional reasons, but also and above all by a special oath sworn to the Reichsführer SS. Moreover, every SS doctor is, as a member of the SS, under an obligation to respect the honour of pregnant women, whether they became pregnant before or after marriage...

Himmler's Order no. 106 specifically forbade the taking of photographs of women received into the homes, and the medical superintendents were made personally responsible for ensuring that this order was obeyed.

Many more orders followed. On 24 May 1937 unmarried mothers were given the right to change their name or that of their child, and were officially entitled to be addressed as 'Frau' instead of 'Fräulein'. On 9 January 1939 they were given the right to withhold the name of the child's father, except in the Lebensborn. On 28 January 1939 an order was issued with immediate effect that birth certificates concerning illegitimate children born in Lebensborn homes could be given only to the mother and the father. They were in no circumstances to be given to officials or private persons or organizations, including party organizations. Should any such certificates be applied for, the application was to be forwarded to headquarters immediately.

Order no. 119 of 5 June 1939, 'concerning illegitimate children born in the Lebensborns', read as follows:

> Following an agreement between the genealogical office of the Reich Minister of the Interior and the L organisation, it is possible to maintain secrecy about the origin of illegitimate children born in Lebensborn homes for an unlimited period. The Reich office will provide a certificate confirming the child's Aryan descent. This certificate can be produced by a child born in a Lebensborn on starting school, for the Hitler Youth and for institutions of higher education without the slightest difficulty arising.

In the event of any breach of the rule of secrecy by any individual or organization, the Lebensborn headquarters had to be informed immediately, and the information had to be passed to Himmler. Offenders were liable to severe penalties. In the files we found a number of documents submitted to headquarters for decision and action. Among the alleged offenders were persons who had made a false declaration of paternity; false witnesses of all kinds; unwed mothers who went back on their decision not to divulge the father's name; young girls awaiting delivery in a Lebensborn home while their parents were looking for them.

The files from which we might have learnt the sequel to these dramas have disappeared. Sometimes a letter that escaped destruction gives a clue, as, for instance, this one of 6 February 1939 addressed to the Gestapo by Dr Wirth, one of the SS medical authorities:

> It is imperative that you take in hand the situation of unmarried mothers. Two girls have committed suicide because they were pregnant. Every means must be used to combat the abortion psychosis.

A baby born in a Lebensborn could be either handed over to become an 'SS child' and a candidate for adoption, or the mother could keep it and bring it up herself. If she opted for the former, she received an allowance on leaving the home for 'post-natal exercises and beauty treatment'.

The greatest care was to be taken over adoptions. Most of the children, after being carefully selected and classified according to their racial value, were offered to people of means. The Lebensborn idea was that the status in life of these families should be as similar as possible to that of the child's natural father (and, if possible, his mother), and that there should also be a corresponding racial and ideological similarity. A child fathered by a senior officer in the police or the SS could not be entrusted to the family of a sergeant-major, still less to that of a private. Knowing all about the father (generally a married man), and knowing that the mother was of 'racial value', the Lebensborn authorities tried to place the child in an environment corresponding as closely as possible to that which would have been his if the circumstances of his birth had been normal.

Childless couples were given priority over 'more fortunate' families, but this principle was not invariably followed. The documents show many instances of children adopted by families in which there were already three, four or even six children. In these cases the father was usually a friend of a senior Lebensborn official.

The documents, and above all conversation with former members of the organization, show how adaptable was Nazi morality. This is what Peter Schütz, the former chief book-keeper at Steinhöring, had to say:

'Let me give you a quite simple example. Colonel Meier had an affair with Fräulein Huber the final outcome of which was a baby. So these homes were established to collect these babies and dispose of them.'

This is what a former secretary had to say:

'Very often the following occurred. A high official in the party or a senior officer in the SS or the army made a girl pregnant. He then arranged for her to have the child in a Lebensborn home. A few weeks later, by agreement with headquarters, he arranged that the child, which had been reserved for him, should be adopted into his own family. He thus killed two birds with one stone. He both produced a child in accordance with the Reichsführer's wishes and at the same time increased the size of his family, which gave the Reichsführer additional pleasure.'

In this connection Peter Schütz told us his own story:

'At that time I had no children, so I was not promoted. When I should have been promoted to commissioned rank, the Reichsführer wrote on my papers in green ink, which was his favourite colour: "No children, promotion impossible" . . . I was a clean member of the Waffen SS, and the Waffen SS was a clean body of men. It was only because of my stomach trouble at the time that I was sent home from Russia and became chief book-keeper at the Lebensborn.

The other kind of SS were the "party hats", as we called them. We had nothing to do with them . . . Herr Sollmann wore the Waffen SS uniform, but he came from the general SS, and wore uniform to which he had no right. Those people are what I call the party-hat SS. They were the upper ten thousand, and it was they that gave the SS such a bad name. And perhaps the whole SS has suffered because they were all put under the same hat.

'Where the adoption of children was concerned, other ranks were automatically at the bottom of the waiting list. The others, the big shots, who were ready to do anything to put themselves in a favourable light, were served first.'

Those girls who preferred to 'take their parcels with them', as the popular saying went at the time, returned to ordinary life as unmarried mothers. The Lebensborn generally forced the father to pay for the child and contribute to the expenses of the mother's stay. If the money was not forthcoming, the organization paid her a small allowance until she was able to manage for herself. In theory, at any rate, no mother or child was allowed to leave until the authorities were satisfied that the mother had adequate means to support the child.

A letter from the Wernigerode home, however, throws a side-light on the aid provided:

Frau Gretel W. is the mother of two children by the same father. She was engaged to him, but he refuses to go through with the marriage. She is in receipt of a maintenance allowance of 30 marks for the first child and 45 for the second, and wants to know whether it might be possible in this instance to increase the allowance for the first child.

Dr Ebner commented as follows:

I have learnt from the head of the home that the father is an Alsatian teacher well known for his anti-German feelings. He is also said to have written to the child's mother that after the war he will probably go abroad. It seems advisable to pass this information to the appropriate Gestapo office.

At the beginning of the century an unknown German

professor said that the worst thing about the Germany of his time was that the women no longer wanted to have children. Himmler's diary comment on this was: 'God grant that this may change.' With the aid of the Lebensborns, it changed to such an extent that Dr Ebner eventually felt justified in claiming that 'thanks to the Lebensborns, in thirty years' time we shall have 600 extra regiments'.

5. Everyday Life

'Nordic humanity made its appearance as the jewel of the earth, the radiant product of the joy of creation. The human being of Nordic race is not only the most gifted but also the most beautiful. . . The harder outlines of the man's face, the gentle woman's face, the live, glowing skin, the bright victorious eyes of both; and common to both the perfect movement of a perfect body! – A regal species among human beings!'

HANS F. K. GÜNTHER, *Ritter, Tod und Teufel*, Munich, 1937

Himmler, whose ambition it was to populate Germany with 120 million Teutons by 1980 at the latest, took a personal interest in the children born with the aid of the various agencies under his control. There are thousands of his god-children in both parts of present-day Germany; nothing gave him greater pleasure than offering his services as god-father, thus ensuring that the child's life began under the best possible auspices. Babies who, in addition to possessing the right kind of antecedents, were patriotic enough to be born on 7 October, his birthday, were the objects of his special solicitude; whole files of correspondence about them survive, announcing the dispatch of toys, embroidered pillows, trumpets, puppet theatres, and the celebrated savings bank book that every child born in a Lebensborn was supposed to possess. This magnanimity explains the bitterness of one Frau Gerda Maria Rothe, who bombarded the Lebensborn head office with complaints and protests when the highly secret registrar's office at Steinhöring II made an appalling and unpardonable mistake. Her child, born on that glorious day of 7 October in 1943, was registered as having been born on the 6th. Fortunately, thanks to Himmler's personal intervention, the error was rectified and the drama had a happy ending.

There were no limits to Himmler's concern for these

children. He intended to take a personal interest in the height and weight of each of them until they reached the age of twenty-one. He gave orders that at birth each should be sent a candlestick, and on every birthday, after it had been weighed and measured, it was to receive from the Reichsführer the sum of one mark and as many candles as it was years old. The first candlesticks, 10000 of them, were made 'gratuitously' for the Lebensborn office in 1940 by the inmates of Dachau concentration camp.

To take the place of the normal christening, Himmler instituted a so-called 'name-giving' ceremony. Frau Ebner described the occasion to us with a great deal of feeling:

'There was always a naming ceremony, and an official from the register office was always there, and the mothers put on their best clothes and smartened up the children, and the living-room – the day room – was decorated with flowers, and it was all very solemn. There were always about ten women, or even more; they held their babies in their arms, and then one by one each baby was given its name . . . and it was always a very enjoyable afternoon, with coffee and cakes, and it was really very nice indeed, with the gramophone records, and so on, it was always most enjoyable. Yes, if you want to call it that . . . it was an SS ceremony.'

However, these ceremonies do not seem to have enjoyed the unqualified approval of all unmarried mothers. There is evidence that up to 1938 some at least had their babies secretly christened by the village priest. To put a stop to this, 'patients' were forbidden to leave the maternity home grounds except with an SS escort.

Himmler's interest in the details of Lebensborn life was so great that he even gave orders that all left-handed children should be trained to use both hands. Also he sometimes commented on women's applications for admission; a 'race expert' having drawn his attention to a 'woman of Slav type but with a slight Nordic admixture, 1·58 metres in height and weighing 58·5 kilogrammes', Himmler noted in his own hand: 'Very good for propagation. Desirable that she should have children.' Even the shape of the mothers' and children's noses interested him. Here is a letter from Dr Brandt to Max Sollmann, the president of the Lebensborn Society, on 15 May 1942:

My dear Max, the Reichsführer SS wants a special card index to be kept of all mothers and parents having a Greek nose or the rudiments of one. As an example of the type required, you should refer to the mother in questionnaire L 6008, Frau I.A., *née* B.

And another, to the RuSHA, dated 13 February 1944.

... The Reichsführer SS, having again gone into the question, has decided that the Race and Settlement Head Office should take steps to draw the Reichsführer's attention to members of the SS and their wives who have Greek noses.

He read all the letters sent to him by mothers in the homes and answered them personally. Also he often had himself photographed with young mothers, to whom he sent a duly signed copy of the photograph immediately on his return to Berlin. He took a great interest in problems of breast-feeding, and came in person to congratulate Frau Anni O. on her record production of 27880 grams of milk in the week beginning 1 January 1940. Good milk producers, mothers who breast-fed their babies for a long time, received awards and were entitled to a prolonged stay either in the Lebensborn home or in a convalescent home where they were the object of every attention.

For mothers' milk too was subject to racial demands. In a letter to Dr Grawitz, head of the Nazi Red Cross, Himmler gave orders that wet-nurses must be of German race; hence the special treatment allotted to these record-breakers. It was also proposed to establish special homes for wet-nurses near the maternity homes. But some women, less patriotic than Frau Anni O., refused to feed their babies, because they wanted 'to keep their figure and the shape of their breasts'.

The correspondence between inmates of the homes and the management provides invaluable insights into their everyday life there, their worries and problems. Some of the letters are so outspoken that one wonders how the prudish Himmler reacted to them. They describe how and how often these 'breeding women' had to submit to the will of philoprogenitive males for the purpose of achieving what was required of them in the Führer's service.

But most of this correspondence deals, with a seriousness sometimes bordering on tragi-comedy, with such subjects as abortions, premature births, false pregnancies, sterility, Caesarean operations, 'unusual' fecundity, wanted babies that did not arrive and unwanted ones that did, 'unknown' fathers who had been sent to the front for breaches of discipline, women complaining and leaving a home 'because the obstetrician is only an ordinary dentist' (the Wienerwald home in Austria). One SS man who had been treated for sterility informed the organization that he had broken a leg just when he was on the point of impregnating [*sic*] his wife. He deeply regretted this 'unfortunate contretemps'.

One girl wrote to the organization: 'I want to give three children to my Führer straight away. What should I do?' The answer was: 'Begin by doing what is necessary to have one.' A letter of 11 August 1941 shows the extent to which fecundity was a duty. In reply to a young woman who asked: 'Can a girl of twenty-three marry a man of forty-eight?' Dr Ebner wrote:

> To my way of thinking this is not normal. Of course the husband should be older than the wife, but there is no sense in this marriage, as the man will be sterile by the time the woman reaches her full maturity. With a young partner a young woman can have many children. Hence our country is not interested in such marriages.

Max Sollmann claims that every effort was always made to persuade the man to marry the girl, but the correspondence between girl aspirants to legitimacy and Dr Ebner tends to contradict this claim. Girls who informed him of their wish to marry the father of their child received the stereotyped reply: 'We are not a matrimonial agency.'

In practice, the Lebensborn authorities adopted different attitudes in different cases. Concerned as they were to establish a high moral tone to counter the flood of ugly rumours that spread about the organization, they tried as far as possible to reply to correspondents in their own terms. On the other hand, here is a letter sent by Dr Ebner on 23 October 1942 to an unmarried mother who wanted the Lebensborn to find her a husband:

I conclude from your letter that you wish to marry only to provide a little brother or sister for your little daughter. It is obvious that it is every woman's heartfelt wish to give her children a father and a home. But this understandable wish can certainly not be fulfilled in every case . . . You as a mature woman will understand me, and I therefore ask you to consider whether your desire for another child could not be fulfilled without getting married.

In this instance Dr Ebner had no need to mince his words, as his correspondent, Frau P. of Berlin-Heinesdorf, had already done her duty in the Führer's service once before.

Other letters seem to lend substance to a rumour widespread in Germany to the effect that many of the Lebensborn's clients were wives who were unfaithful to their husbands while the latter were away on active service and took advantage of the opportunity to have their illegitimate children in secret. There are letters from girls full of naïve protests at the 'incomprehensible' behaviour of the fathers of their babies. Dr Ebner's replies, full of comforting, consolatory advice, fill at least three files.

There were also big differences in the treatment of women in the homes. Some were literally thrown out only a few weeks after confinement, while others remained for months, or even a year or more. It is impossible to discover any basis for these differences. Apart from any personal or racial considerations, shortage of beds probably played its part, for during the heyday of the institution between 1940 and 1944 the homes were working all out, and at some there were more than five hundred deliveries yearly. This superabundance of clients led to a phenomenal amount of coming and going between the homes. Women whose names were down for Steinhöring found themselves in Belgium when the time came, and after the *Anschluss* some had crossed the whole of Germany to have their babies in Austria. There were also women from Occupied France, Belgium and Holland who used the Lebensborns to have their SS boy friends' babies and hand them over to the SS.

The honour of being a German mother, and the obvious material advantages, not the least of which were excellent food and comfortable accommodation, were, however, accompanied by certain minor disadvantages. In the first place, Christian names only were used, to avoid any discrimination between married and unmarried women and to preserve their anonymity. Also every inmate had to clean her own room. These requirements caused dissatisfaction among wives of senior officers in the SS, the police or the party. They resented the presence of girls or women whose social and educational levels were, they felt, below their own. Letters from future mothers who complained about 'the atmosphere, the food, the doctors or the nurses' were numerous, and their complaints long and detailed. Officers' wives refused to be addressed in the familiar second person singular, and insisted on their Christian name being preceded by 'Frau'. Unlike the unmarried girls who flocked there as soon as their condition became visible, they were married; moreover, they were married to dignitaries of the régime and failed to see any reason to conceal their identity. Some refused to stay a single night after the delivery of their babies, which explains the numerous reports of sudden departures.

Concerning the wider fields of cleanliness, hygiene, discipline and diet, numerous reports, orders and instructions give a very clear idea of the state of affairs existing in this or that home at a particular time. The files show that total chaos often prevailed in those homes in which the future élite of the New Black Order of the SS were to be born. Among the things mentioned are bits of wire or nails in the babies' broth; chamber-pots filled to overflowing but not emptied for days or even weeks; infants kicked by nurses while seated on these pots; indiscipline on the occasion of the 'enthronement' of a new head of a Lebensborn home.

when it was necessary to stand to attention, giving the Hitler salute, even if the ceremony lasted for hours, a thing that was too much to ask 'in view of our condition'; reprimands given to a pregnant woman 'who sang at the top of her voice' (this was reported to Ebner, whose comment was: 'Singing is not harmful during pregnancy'); heart-rending complaints by women to whom the wrong baby had been returned by mistake.

At the beginning of January 1941 an unexpected visit by Himmler to the Kurmark home at Klosterheide, near Berlin, made a most unfortunate impression on him of disorder, neglect and total indiscipline. A few days later, on 11 January at 1300 hours precisely, he summoned a meeting of his principal subordinates, including General Karl Wolff, Dr Ebner and Oswald Pohl, who was appointed chief administrator of the concentration camps the next year, and executed as a war criminal after the war. The minutes of the meeting are revealing.

> The Reichsführer SS also inspected some wardrobes at the Kurmark home and in some cases noted the greatest disorder. The mothers are to be trained in tidiness. Wardrobes must not be locked. In the event of theft, the thief must be severely punished.
>
> Cases of severe indiscipline by mothers should be punished by immediate expulsion. It will . . . be advisable to expel a mother about every six months as an example. This will undoubtedly have a beneficial effect on the others.
>
> Men visitors to the homes are to be forbidden in future. The homes run the risk of losing their good reputation through such visits. To avoid hardship, a so-called visitors' hut will be set up in which visitors could perhaps be given coffee, but in which no possibility of intimacy will exist. . .
>
> Two expectant mothers whose children were fathered by the same man were in the Kurmark home at the same time . . . There must be an investigation into this, and such coincidences must be avoided in future . .
>
> The head nurse at the Kurmark home, Erika, has asked to be allowed to adopt a child to whom she has become particularly attached. This request is to be granted.
>
> The Reichsführer SS intends to have an artistic statue representing a mother and child placed in an appropriate situation

outside every home to indicate its nature. The statue of a mother feeding her child that has been sent to Steinhöring is to be set up at an appropriate spot immediately.

Each home was guarded by an SS detachment, and the women had to show their passes before leaving the grounds. The primary task of the guards was to keep inquisitive eyes away and check the comings and going of the inmates.

Women who worked outside sometimes picked up infections which spread to the whole establishment. These worked such havoc that at some homes the women were eventually forbidden to go out. If a woman had to go out with her child for some valid reason, both were put in quarantine for a week after their return.

Sometimes the homes became the scenes of the most ridiculous incidents. At Wienerwald, near Vienna, in 1939 a woman, a professional midwife who had already had two illegitimate children herself, was expecting a third. Although male visitors were forbidden outside visiting hours and were admitted only to the crèche, where a mother's other children were kept during the final months of her pregnancy, this future mother had her 'fiancé' in her own room for a whole night. She was threatened with expulsion, but, unfortunately for Dr Ebner, the man was not only an SS officer but also a personal friend of Himmler's. He threatened to report the affair to the Reichsführer, because 'a mother of two children had been insulted'. At this point he went down with a terrible attack of rheumatism, which had to be treated on the spot, with the result that he spent several weeks in the home, his presence making the other girls hysterical. But as soon as he had recovered sufficiently, the row continued. Ebner insisted that if he wanted to see his children he must come during the regular visiting hours and not at night. The man replied that he refused to allow the mother of his children to be treated as a semi-streetwalker. Ebner had the last word, however. The midwife caught whooping cough and was sent away 'to avoid the risk of an epidemic'. But the story had a happy ending. The couple eventually set up home together in Vienna, where the husband still exercises his profession as a psychiatrist.

Indiscipline was not restricted to such intimate matters. Some mothers talked about things that were regarded as unacceptable by the Lebensborn authorities, not because of the facts revealed, but because they were considered unsuitable in such an environment. This letter to Sollmann from the head of the Wienerwald home, dated 16 April 1943, provides an example:

... since the beginning of her stay here [Agnes Spangenberg] has caused considerable unrest among the mothers by gossiping about her work with the Gestapo in Smolensk. She gave detailed descriptions of mass executions of Jews, where babies are also said to have been killed by a shot in the back of the neck. I have vigorously reprimanded her, as stories of this kind are certainly not suitable in the setting of a maternity home.

Sollmann's comment on this letter was: 'I think Dr Schwab deserves a letter of commendation in this case.'

Wholesale waste and theft prevailed at all the Lebensborn homes. Head nurses were dismissed literally for plundering. Ebner himself had to 'dismiss the head nurse Anna H. for stealing porcelain, soap, linen and other things'.

A large number of letters, orders and circulars both in Himmler's personal files and those of the Lebensborns testify to theft, fraud and blackmail of various kinds practised by persons in positions of responsibility in the homes. After the war, according to evidence given at the Lebensborn trial at Nuremberg in 1948, their rooms were found full of fur coats, silk stockings, jewellery, bottles of champagne – and children's clothing.

The bad reputation of the Lebensborns was based on the luxury, waste and extravagance that prevailed, combined with the secrecy that surrounded them. The files we consulted at Arolsen are full of edifying information about the popular attitude towards 'those girls that don't work and spend their time titivating themselves up and showing themselves off in the streets'. The term 'whore' continually recurs in this connection.

On 22 December 1941 a consignment of 68·25 kilogrammes of sweets and chocolates arrived at Ebersberg station, addressed to the home at Steinhöring. The news spread in the village, and some anonymous letters of protest arrived. But the villagers were afraid of reprisals, all the principal staff being members of the SS.

In another village, near Wiesbaden, some Lebensborn girls were beaten up while taking the air in the evening. Towards the end of the war other ranks serving with the Waffen SS wrote to the Lebensborn organization asking that their wives and children be sent something to eat. The answer, signed by Ebner, was always the same: 'I regret being unable to help you in your situation and greet you with Heil Hitler.'

A questionnaire sent to the homes on 20 April 1942 shows how the war and Germany's first reverses failed to put the Lebensborns and their everyday problems, especially those of diet, out of Himmler's mind.

> How is the cooking done in the homes? Is attention paid to the correct steaming of vegetables so that their nutritive value is not lost?
>
> Are those absurd boiled potatoes still cooked, or is strict care taken to ensure that potatoes are served only in their skins?
>
> Is sufficient uncooked food provided, e.g. raw sauerkraut or raw carrots?

Another of Himmler's circulars states:

> As far as possible wholemeal bread will be eaten in all Lebensborn homes.
>
> The Reichsführer SS asks the heads and head nurses of the homes to ensure that porridge in a sufficiently appetizing form is served for breakfast in the homes . . . The Reichsführer described as absurd the fear that women might lose their slim figures by eating porridge.

Himmler was greatly concerned with the question of porridge. He ate it himself because he suffered from terrible stomach-aches, for which he was constantly treated by his

Finnish masseur, Kirsten; he also consulted his old friend Dr Ebner. Like most of the leaders of the Third Reich, he had a severe inferiority complex towards the British, 'those porridge-eaters who nevertheless stay thin'.

> Just look at Lord Halifax. His slender figure shows that those oatmeal flakes they called porridge have no influence on the weight of persons of quality . . . the Reichsführer wants mothers and staff to get used to porridge for breakfast and later to introduce their families to it.

So far as vitamins were concerned, a woman doctor at the Sonnenwiese children's home reassured the Reichsführer in the name of her colleagues:

> All potatoes are cooked in their skins, two-thirds are boiled, and the rest are roasted or mashed. Raw food is provided in the form of fruit, and raw carrots are served, partly in salads. Other vegetables suitable for the purpose are unfortunately not obtainable at the present time. Heil Hitler!

Not only were the Lebensborn inmates treated to porridge for breakfast after this, but throughout the war they never went short of real coffee, tea, cod-liver oil or fresh fruit. Towards the end of 1943 a barrel of 200 kilograms of cod-liver oil turned up accidentally at the concentration camp of Dachau, only a few miles away from Munich and Steinhöring. After the usual correspondence, orders were given to distribute the contents to the Lebensborn homes in the north and south of Germany as 'the camp inmates have no need of its strength-giving qualities'.

Some former nurses and head office employees whom we questioned in 1973 recalled with a great deal of nostalgia 'the marvellous food served in the homes right up to the very end of the war'.

Frau Maria, who was a secretary in the Lebensborn registrar's department and now runs a small hotel in Bavaria, told us:

'Even during the war we always had fresh fruit, it arrived in truck-loads. We always ate well. It was only after the war that it deteriorated. During the war the children thrived, and all of us, whether members of the staff, mothers or children, always had plenty of excellent food. It was very varied, though often of course the menu was repeated from one week to another. But it was certainly very good. I still remember when fresh cucumbers arrived. At the time we had a head nurse who had those fresh cucumbers served up as vegetables instead of as salad; I didn't like them served up as vegetables, and I was sorry about those lovely cucumbers that I should have preferred to have eaten as salad. I remember that very well, but otherwise the food was perfect.'

Frau Maria added: 'We never went short of anything. It was a marvellous time, the best time of my life.'

6. Footing the Bill

> 'What is at stake in our struggle is National Socialism, an ideology based on the value of our Teutonic blood, our Nordic race. What is at stake is a world as we have planned it, a fine, honourable world based on social equality, a world rich in pleasures and culture.'
>
> HIMMLER at Stettin, 13 July 1941

The development of the Lebensborn organization was stimulated first by the SS, then by the war, and the long succession of German victories. The result was a fantastic growth rate, so that between 1940 and 1944 it employed several thousand people and hundreds of millions of marks had been spent.

In the early days of the SS and its various organizations, the racial policy of the régime was carried out thanks to the generous aid of German industry and high finance. Bankers, industrialists and members of the aristocracy paid their contribution to creating the new age of promised Teutonic supremacy. Among the most generous contributors within the circle of 'the Reichsführer's friends' were IG Farben, Siemens, Krupp, the Dresdner Bank, the Reichsbank, the Deutsche Bank and the food, oil and pharmaceutical industries. For obvious reasons, most of the evidence has vanished from the files, but Himmler's correspondence, which has survived almost intact, contains copies of personal letters of thanks to 'big shots' in Berlin, Vienna and Hamburg. It is curious to note that, though there were innumerable other SS organizations, in his replies he almost systematically emphasized the importance he attached to the Lebensborn organization.

13 March, 1939. SS Gruppenführer Wolff to SS Gruppenführer Ernest Kaltenbrunner.

Dear Kaltenbrunner,

SS Oberführer Cassel [Kaltenbrunner's chief of staff] was asked by telex on 5 August to pass on to SS Gauleiter Dr Jury [Gauleiter of the Lower Danube] the Reichsführer SS's instructions that the licence for the Baden–Wienerwald casino is to be granted only on condition that the surplus net profit goes to the Lebensborn Society . . .

The Reichsführer SS is convinced that Gauleiter Dr Jury will fall in with his wishes. . .

The Lebensborn, like all other Nazi organizations – though perhaps even more, because of its special role in building up and fortifying the race – was chiefly financed by money and property expropriated from Jews.

The first confiscation of 'enemy property' yielded 93 366 358 marks, which was equitably shared between the Nazi Party, the Wehrmacht, the SS and the Lebensborn organization, among others. But the buildings, property, art objects, medical installations and so forth worth that amount represent only a minute fraction of the 'property taken from the enemy' – in other words, from the Rothschilds and Habsburgs and from religious orders and Jewish communities in Germany and Austria. And this was the sum raised in 1938 alone; to it must be added hundreds of millions more subscribed, offered and stolen in the name of the master race.

In the light of the principle of preserving Aryan life and destroying 'sub-human' life, the role of the Lebensborn 'charitable society' becomes clear. The Jews whose money financed Himmler's Aryan stock-breeding were simply thrown out of their clinics and hospitals. Doctors, professors of medicine, administrators, nurses, forced to adopt the additional forenames of Israel or Sarah, were deprived of their rights by the Nuremberg Laws and sent to concentration camps.

The other 'enemies' of Germany – the Churches and a few rich 'foreigners' who refused allegiance to the swastika – fell under the ban too, and their wealth also contributed to swelling the bank accounts of the SS, but their fate was in no way comparable to that of the Children of Israel. The Nazis confiscated the property of religious institutions without sending priests or nuns to concentration camps, unless, of course, they showed 'hostility' to the régime, which, in the context with which we are dealing, was rare.

Similarly, the foreign 'enemies', who in any case were few, at most risked expulsion.

In the distribution of this booty, which took place throughout Europe at the expense of one and the same minority, the Lebensborn organization was a major beneficiary. The place of those about to die by the Führer's will was to be taken by those who were to be born by his will.

In 1938 things seemed to be going according to plan. When the Wehrmacht marched triumphantly into Austria, the chief administrators of the Lebensborns, Pflaum, Ebner, *et al.*, followed closely on its heels; and while Hitler was addressing 100000 people from the balcony of the Imperial Hotel in Vienna, Pflaum and Ebner were feverishly consulting the telephone directory of Vienna and the surrounding area. That same day they picked out the Neulengbach children's home and the Wienerwald sanatorium; the former belonged to a Jewish charitable organization, the latter to two Jewish doctors. The Lebensborn organization ensured that these people were 'taken care of' by the Gestapo. One of the doctors committed suicide and the other died in a concentration camp.

The choice of these institutions was not left to chance, and they were not confiscated simply because they belonged to Jews or religious organizations. Ebner, aided by some officers, systematically inspected, photographed, took measurements, demanded information, made plans and calculations, though he had received specific instructions that expense was no object. 'Two or three million marks more or less was immaterial,' Sollmann testified at his trial later.

After weeks or months of discussion, the best sites were picked without regard to price; there had to be plenty of fresh air and plenty of room. Villas, sanatoria, hospitals, private clinics and children's homes, the occupants of which were sent away for 'disinfection' or 'special treatment', were largely reconstructed, internally and externally. High walls were built and rows of trees planted to shelter them from prying eyes. All this was very expensive, of course, though labour was cheap, being available from the concentration camps in the area. In fact, innumerable concentration camp inmates – including Frenchmen, Poles, Dutchmen and some German Jehovah's Witnesses – were employed on heavy labour at the Lebensborns. Frau Ebner claimed that all such were very well treated at Steinhöring, and said she remembered in particular one Frenchman who suffered from indigestion, to whom they often gave white bread. That Frenchman, Pierre Variot, a member of the resistance who was caught and sent to Dachau, where he became a member of the 'Steinhöring gang', today lives at Lyons. His recollection is that the Lebensborn people treated the deportees harshly and inhumanly. All were afraid of Sollmann, and particularly of Dr Ebner, who used to watch them working in the garden through binoculars. They were threatened with severe punishment if they entered the home, and were forbidden all contact with the residents. A number of former concentration camp inmates testified at the Nuremberg trial that they had been badly treated by the Lebensborn people, including Sollmann.

From the start the Lebensborn organization was administered from Munich; the choice of that city was no doubt connected with the origin of its leading figures, who, with few exceptions, were Bavarians. Ebner was born at Ichenhausen, near Munich, Sollmann at Bayreuth, and Himmler in Munich itself. The organization requisitioned the headquarters of the Jewish community in the Herzog-Max-Strasse and made it its own headquarters, where it employed a staff of several hundreds. It established a number of maternity homes in Munich and the surrounding area, as well as children's homes, convalescent homes and even two old people's home for members of the SS. It also established the first real Lebensborn home at Steinhöring. Among the premises it requisitioned was the house in the Poschingerstrasse that belonged to Thomas Mann, and in his post-war memoirs the novelist wrote at length about the Lebensborns and the selection of Nordic girls practised there. The information came from his housekeeper, whom the SS retained. She described the medico-racial examination to which the girls were subjected with a view to their producing children for the Lebensborn organization, and supplied many details, particularly about the measurements taken, which ranged from the shape of the girls' heads to the width of their pelvises.

At his trial Sollmann admitted that after he took over in 1940 two other hospitals were 'transferred' to the organization, one in Munich and the other at Nordrach in the Black Forest, as well as two old people's homes, a children's home in Munich, several villas at Karlsbad (Czechoslovakia) and in Vienna, and about a dozen villas and flats in Munich. All these had belonged to Jews, but, as Sollmann naïvely declared, they had fallen vacant. He also mentioned the huge sums that flowed into the Lebensborn coffers in different forms and from different sources; the one thing these had in common was that they came from 'enemies of the Reich', Jews in particular.

After the war the Americans were surprised to discover that 20 million marks were still credited to the 'L account' at the Münchner Bank, to say nothing of smaller amounts in dollars, francs, guilders and zlotys. What these amounted to in all was never established, but Sollmann's estimate was 20 million marks. In comparison with everything else, the allotment to husbandless mothers of free accommodation in premises which belonged to 'absent' Jews was a mere trifle; Sollmann and Ebner had personally intervened with the authorities to help their clients over their temporary difficulties by securing them this accommodation.

The more Jews were arrested, the longer grew the list of property 'entrusted' to the Lebensborn organization. In 1940 Hitler personally signed an order transferring to it a large sum of money 'withdrawn' from various Jewish bank accounts. German railway receipts survived the accidental or deliberate destruction of the records of the 'L office' from areas where death, whether slow or sudden, prevailed. Those loads of clothing, provisions, furniture (beds, tables, chairs, cupboards, crockery, and so on) came from Poznan, Cracow, Bialystok, Lodz, Chelmo and Warsaw, and later from Kiev and the whole of German-occupied Russia.

A Lebensborn file contains the following: 'Large consignment of linen, etc., received this morning from Cracow.

Sent by SS Kruger.' Another extract reads: 'Consignment of clothing sent from Poznan to Lebensborn Munich. Copy of dispatch note enclosed. Heil Hitler.' Letters or telegrams of this kind flowed in throughout the war, chiefly from the Eastern occupied territories, but also from Holland, Belgium, Luxembourg and France. The Lebensborn organization took over the Château Ménier at Lamorlaye, near Chantilly, in France, a Jewish sanatorium in Holland which became the Gelderland home, and a number of clinics, hospitals and maternity homes in Belgium and Luxembourg (to be dealt with in a later chapter). In France, Belgium and Luxembourg, requisitioning was done in Wehrmacht style, that is, the occupants were simply ordered out and the German command took over. But, in Holland, these 'mopping-up operations' were carried out in classic SS style. Psychiatric hospitals were stormed and plundered and the occupants were maltreated and sent off in train-loads to Auschwitz. None of the patients, whether adults or children, or the staff who accompanied them ever returned; all were gassed. All the patients and staff of Apeldoorn, a large mental hospital, died at Auschwitz or Treblinka by order of the German Dr Kroll, who handed them over to the SS. Non-transportable patients were shot on the spot. A few weeks later the first Aryan baby was born in one of the sanatoria, repainted and decorated for the occasion. The happy papa was an SS Unterscharführer (sergeant), and the mother a Dutch girl of nineteen and a half.

The following is a letter from Dr Ebner, chief medical superintendent of the Lebensborn homes, to Max Sollmann, the chief administrator. He considered it so important that he had it delivered to Sollmann personally by a member of his staff.

Proposal for the use of the installations at Otwock (Poland).

The buildings in O. near Warsaw . . . are admirably suited to be L homes. Because of their layout and construction they can be described as good, well above the average Warsaw standard. In addition, the climatic conditions at O., where until now there were tuberculosis sanatoria, are exceptionally favourable. Practically no reconstruction is required in either of the buildings. The chief requirement is the repainting of all rooms, corridors, doors and windows. The medical equipment, linen, and part of the furniture can be procured from the Warsaw ghetto with the consent . . . of the General Government.*

This letter is dated 10 February 1943, when terror, starvation and death had reached a high point. In the Warsaw ghetto 60000 Jews were confined in conditions of indescribable horror to an area of 3500 square yards. It was the hardest winter of the war in Poland. There was no heating, no water, no bread, nothing. The Jews, faced with extermination, had revolted on 18 January, only four days after a surprise visit from Himmler accompanied, among others, by several senior officials of the Lebensborn organization, who came specially to assess the amount of loot to be expected in the ghetto.

One of them was Dr Ebner, who was particularly interested in the '100 mothers and children who might be accommodated in House B1' of the Otwock complex, which was later to become the Ostland maternity home. He continued the above letter:

I therefore propose that the well-equipped sanatorium of the city of Warsaw be cleared of tubercular Poles, who should be accommodated in a suitable hutted camp,† and that the vacant institution be used for tubercular patients of German stock. The public health authorities of the General Government are hardly interested in restoring tubercular Poles to health so long as German tubercular patients are not properly accommodated. . . In House Z4 children suitable for Germanization could be accommodated and tested for a year as to their aptitude and quality to see whether they could be adopted through the mediation of L. . . Thus Otwock, run by the SS in the person of the SS medical officer, would become a model institution in the hands of the General Government.

*The government of rump Poland headed by Hans Frank, another major war criminal, who was executed in Poland after the war.

†Ebner was of course familiar with the secret code for genocide, and the phrase used here simply means liquidation.

However, the Lebensborn authorities were cheated of the ghetto loot on which they had cast covetous eyes that February. It vanished with the destruction of the ghetto and the liquidation of 60000 Jews after a battle that lasted thirty-three days.

7. Maternity Homes or Stud-Farms?

'Top secret, very urgent. To all Reich medical services. Ref. enemy propaganda.
'Rumours easily arise in wartime which are sometimes distorted, sometimes exaggerated, but are sometimes nevertheless circulated in that form. Rumours of this kind often refer to the medical services. We continually insist on the danger involved in spreading these rumours, as well as information communicated in the course of duty that is not intended for the public. Officers and non-commissioned officers are called on strictly to abstain from spreading such rumours.'

Circular from the German High Command, copy addressed to Dr Ebner, September 1943

Thirty years after the collapse of the Third Reich the Lebensborns remain an embarrassing topic of conversation in Germany. Most Germans who lived under the Nazi régime shrug their shoulders and refuse to answer questions, and the minority who are willing to talk generally react in stereotyped fashion by blaming everything on the SS. Just as they lay all the responsibility for the atrocities of the concentration camps squarely on the shoulders of Himmler and the SS, so they hold the Lebensborn organization responsible for forcing German girls to mate with selected males in human stud-farms for the sole purpose of giving a child to the Führer. The attitude of the post-war generation is merely one of amused interest, periodically revived by sex magazines for which some of the innovations of the Third Reich provide an inexhaustible source of raw material.

Soon after its foundation the Lebensborn Society set up its own register office, known as Steinhöring II. This was a personal decision of Himmler's, after the local register office had committed an appalling blunder. Not being familiar with the new methods of the SS in the matter of selective breeding, it had followed what had been the normal practice for generations, and forwarded an illegitimate child's birth certificate to the register office of the mother's home town. The result was a tremendous local scandal, because the baby's father, an SS senior officer, was a married man.

The setting up of a special register office for babies born at the Steinhöring home increased the uneasiness the villagers had always felt about it. The inmates, who had no contact with the local population, arrived several months before their confinement, and spent their time going for country walks. Their idleness shocked the peasants, who still talk about the institution as a foreign body in their midst.

In other parts of Germany where these homes were established the reaction was similar. Why were these special homes for unmarried mothers set up in the first place? Were the existing maternity homes not medically satisfactory? Why should girls and women from Berlin travel several hundred miles by train or car to have their babies in Bavaria, or the other way about? People were intrigued by these new institutions placed in the heart of the countryside and carefully sheltered from inquisitive eyes. To practising Christians, such as nearly all German peasants were, nothing could have been more shocking than the very idea of the Lebensborns, which rejected out of hand the principles of the Church on marriage and baptism.

But what was more disquieting than anything else, at any rate at first, was the secrecy with which the organization was surrounded. Even members of the SS failed to appreciate the way in which their Reichsführer's ideas were being put into practice, as is shown by this memorandum addressed to a colleague by Dr Ebner:

> With reference to the withdrawal from membership of the Lebensborn Society of various SS sub-leaders [in other words, they had stopped paying their subscription] . . . there is a general impression that there is something sinister and mysterious about the organization. At all events, the continual references to secrecy in the prospectuses do not make a good impression on married women. It would be better to remove these references from the prospectuses for the sake of these married women.

People also failed to understand why children born of extramarital relations under the protection of the SS flag were legitimate in the eyes of the law, while others, born in ordinary maternity homes, continued to be regarded as illegitimate. They also found it difficult to understand why the Lebensborn Society, officially declared a national institution, should also be described as a charitable organization. Nor was it easy to see why a charitable organization which advertised the special privileges it enjoyed in its prospectuses should be surrounded by such secrecy.

It began to be murmured that it was really a highly secret SS organization founded by Himmler for the purpose of producing blond, blue-eyed children by mating men and women selected from the best Nordic specimens in Germany; and this rumour, of course, gained steadily in strength as more and more homes were established and the Munich headquarters expanded. Very soon everyone knew someone who had heard of a girl who was going to give a child to the Führer with the Lebensborn's aid.*

The young men 'introduced into these breeding establishments' were popularly known as the SS's 'stud-bulls', but the most insulting terms were reserved for the girls. One claim was that they even refused all medical aid that might have diminished the pains of childbirth, preferring to gaze at the Führer's portrait.

On 27 June 1943 Lance-Corporal Rudolf Müller of the Wehrmacht wrote to the RuSHA saying that many rumours were being spread about the Lebensborns 'which are not compatible with the National Socialist view of marriage and the family'. He naïvely continued:

As my brother, SS Hauptsturmführer Hans Müller, who gave his life on the Eastern Front, was a member of the Lebensborn for several years, I request an explanation of the organization's real purpose, so that I may play my part in putting things right.

*This, we gathered, is the impression still generally prevalent in Germany. The press in other countries carried reports of 'human stud-farms', e.g. *Voice of Freedom*, May–June 1943; *New York Herald Tribune*, 11 July 1943; *The Times*, 14 December 1943. Each gave detailed accounts of 'human breeding' by the Nazis and of 'planned reproduction' from Poland to Norway.

Perhaps an article in the *Schwarze Korps* would be useful. Heil Hitler!

Ebner underlined the words 'organization's real purpose' and angrily wrote 'NO:' in capital letters in the margin opposite the last line.

However, the idea that active sex took place at the Lebensborns made its way even into the higher ranks of SS society. One day Ebner received a letter from an SS Unterscharführer Greber, who asked him to help one of his subordinates, a wealthy wine merchant from the Freiburg area, whose wife could not have children after an operation.

On that estate there ought to be at least two to four healthy children. SS Unterscharführer Rieflin firmly believes that at least half these children should be of his own blood. In view of the fact that his wife's capability and other characteristics put divorce out of the question, SS Unterscharführer Rieflin wishes me to inquire on his behalf whether there might not be a German woman who would agree to give him these children in the present time of destiny.

Obviously he would be willing to provide the woman with an appropriate maintenance allowance before the confinement and also for a certain time afterwards; ... It should also be mentioned that Frau Rieflin is in agreement with this plan and is prepared to give the children a conscientious upbringing.

Ebner noted at the bottom of this letter in his large and regular hand: 'No, we can't find him a partner, but we can give him every other kind of help. (Secrecy).' A fortnight later he answered SS Unterscharführer Greber, but without beginning with the usual 'Dear SS Comrade'. After saying that he appreciated the situation of the Rieflin family, he went on to deal very sharply with the point that obviously upset him – the implication that sexual partners could be procured through the Lebensborn organization:

The whole thing always breaks down when the childless husband finds a woman who agrees to his wishes and the mother does not keep her word after the birth of the child.

The Lebensborns can do nothing in this respect, because the search for and introduction to a suitable partner is such a delicate and difficult business that in the end it must be taken care of by the male partner himself. Heil Hitler!

In June 1943 a Dr E. Brandenburg, a medical officer in the Luftwaffe, was transferred to a Lebensborn. 'No sooner had I arrived than the head nurse Luise W. pestered me with her advances,' he said in evidence at a Munich trial in 1950. He had not yet been initiated into the secrets of the 'L office', was not sexually interested in Sister W., and, being more military-minded than Nazi, told her things about the progress of the war that she regarded as defeatist. Repulsed and disappointed, she got her own back by reporting what he said to Dr Ebner, and a lively altercation between the two took place the same evening. 'I will do what is necessary,' Ebner said, 'to ensure that you, a democrat and a liberal, lose your liberty as quickly as possible. I shall inform Himmler in person. It may cost you your life.'

Soon afterwards Brandenburg was summoned to Berlin to face an SS judge, who said to him: 'Why didn't you fall in with the woman's advances? It's the usual thing in the Lebensborns. You'd have spared yourself a great deal of trouble, and you'd have been able to live in peace.' To another nurse, who was looking out of the window, Dr Brandenburg had apparently said: 'Don't look up in the sky, that's British territory.' In November 1944, he was arrested and deprived of his rank, and the end of the war found him in prison at Wolfenbüttel.

SS Brigadeführer Woss was obviously better informed than Dr Brandenburg, because when his daughter, who was looking for a job after qualifying as a doctor, was offered a position in a Lebensborn maternity home, he diplomatically informed the Lebensborn authorities that she had found a job in an SS hospital and was no longer available.

Annie E., who was eighteen in 1943, had heard that the Lebensborns were looking for nurses and applied to Dr Ebner for a job. The latter replied that he was willing to employ her, but first wanted some personal information, as well as a photograph, which the delighted Annie sent on 14 September. But next day, before he had time to answer, she

Front page of the articles of association of the Lebensborn Registered Society.

The sight of a blonde child softens Himmler's heart.

The SS maternity homes were situated far from prying eyes in the dępths of the country.

Political indoctrination began at a tender age.

The naming ceremony took the place of baptism.

1 The chief administrator.

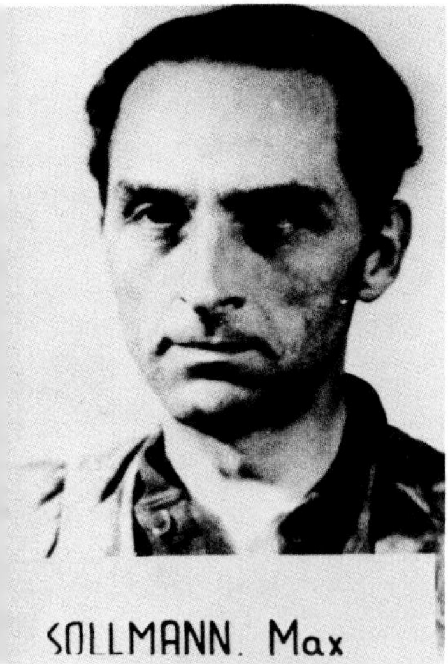

2 The chief medical superintendent.

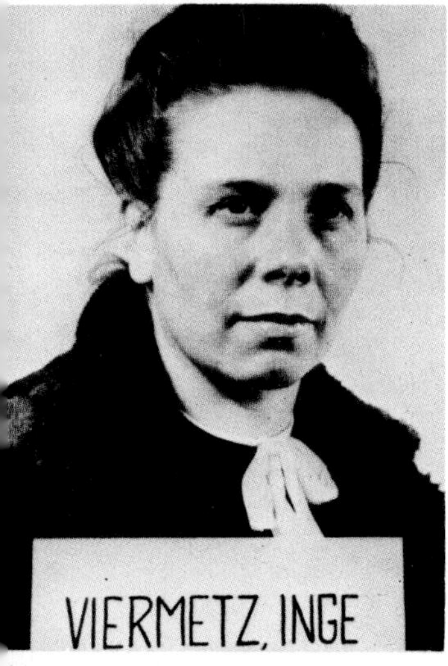

3 Head of Lebensborn homes abroad.

4 Head of the legal department.

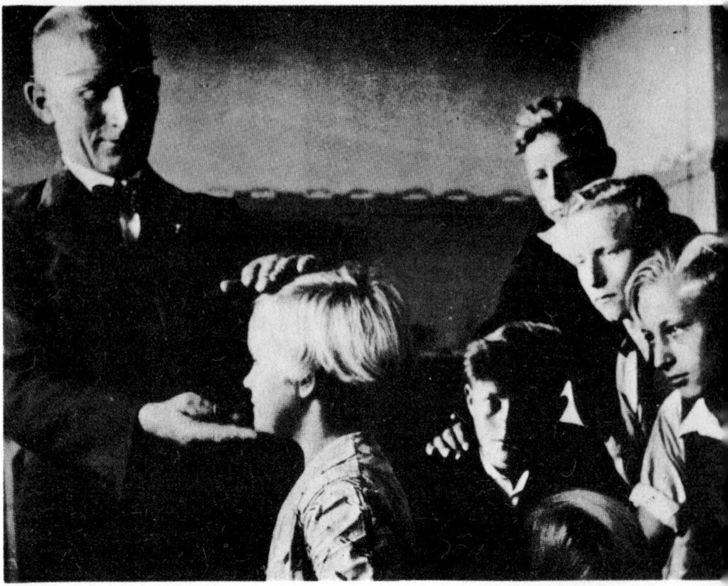

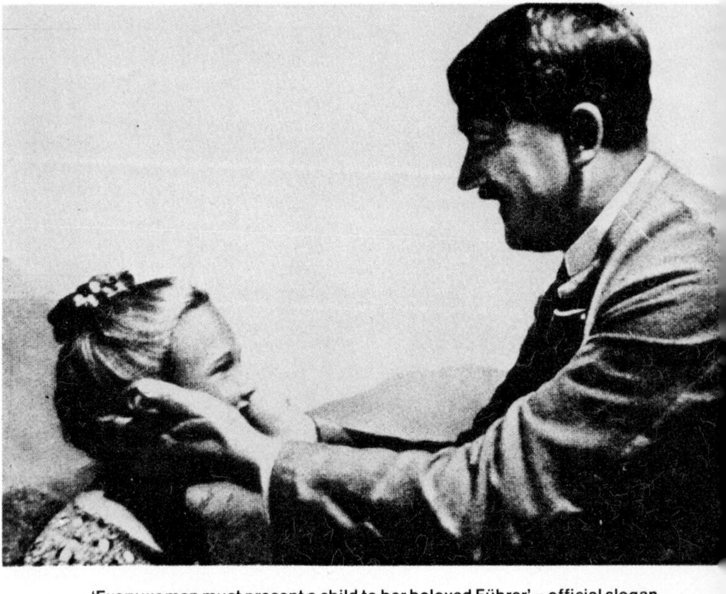

'Every woman must present a child to her beloved Führer' – official slogan.

Above
'Can the same mind, the same soul, inhabit such different bodies?'

Opposite top
Children were encouraged at school to measure each other from the racial point of view. Teachers held up the best Nordic specimens as examples.

Right
Three 'Little Brown Sisters.'

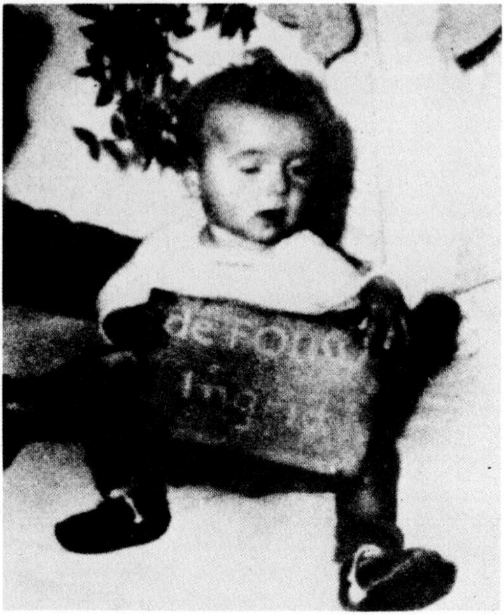

'When I was a child my name was Ingrid de Fouw.'

Below
Ingrid, thirty years later, in front of the former Lebensborn where she was born.

informed him that for health reasons she unfortunately would be unable to accept the job. What had happened to make her change her mind overnight? When we interviewed her near Linz in Austria in March 1973, she told us. 'A girl friend whom I told about my plan explained things to me. And as I did not want children without being married, I used that pretext to get out of it. You understand . . .?'

The most effective recruiting for the Lebensborns was thus done by word of mouth. Those who accepted what went on in them committed themselves completely. Others, like Annie, who for one reason or another found the idea 'hair-raising, disgusting and immoral', tried to extricate themselves as best they could. This was not without its risks, for, having been let into a jealously guarded secret, they could well have found themselves caught in the coils of the Gestapo.

In all the correspondence of the Lebensborn mothers, there is no hint of love or even affection for the men whose children they had just borne, no mention of any sentimental problem. Nor is there any reference to pleasure or sexuality. More than three-quarters of the girls came from the BDM or the Reich Labour Service. Because of their age (fifteen to sixteen) and their inexperience, these solidly indoctrinated guinea-pigs were the most receptive to physical and psychological propaganda for reproduction in the national interest. Sexual intercourse as a duty to the State had nothing whatever to do with sexual intercourse as a pleasure; it was the brick that every German male or female had to contribute to the construction of the edifice. To the Nazi biologists, the love of man and woman and its consummation in the sex act were mere Judaeo-Christian devices for enslaving man to his senses.

In reality, the Lebensborn maternity homes were not so much institutions for women as institutions for men of the SS. A young unmarried SS man, a soldier devoted body and soul to Nazism, capable of destroying anything that might stand in the way of the future Teutonic empire without reflection, compunction or complaint, must not be burdened with the worries of the ordinary man, who might be forced into marriage by a woman's unwanted pregnancy. As for married members of the SS, whom Himmler had ordered to have at least four children – or, better still, four sons, 'for only an SS man who has secured his succession can die happy' – they must be relieved of the cares of bringing up children and of all other moral or material responsibilities. The bastard born of an SS man and a selected woman would have the SS State of which Himmler dreamt for a father and the Lebensborn for a mother. Moreover, the State and the Lebensborn organization would play the part of subsidiary parents to legitimate children born of conventional marriages. With an organization such as that behind it, the SS élite, to whom Hitler had handed the 'blood flag' in July 1926, would be able to dedicate itself to the cult of its Nordic, military, male superiority completely untrammelled – by sentimental considerations in particular. Man's task was to make war, woman's to make children.

Thus the single SS man had complete freedom of action, subject only to the exigencies of his racial code, while his married counterpart, having demonstrated his fertility in the marriage bed, was authorized to indulge his philoprogenitive inclinations outside it. As for the women who were the conscious or unconscious victims of all this, they were advised 'not to force marriage on a child's father, for a forced marriage is not a happy one'. They were also advised 'not to divorce a husband for the sole purpose of giving their child a father, because the Lebensborn would take charge of it'. Thus the SS man had his hands entirely free.

On 21 July 1939 a future father wrote to a girl in Graz:

> When you say in your letter that it would be a great disgrace to bring an illegitimate child into the world, I must remind you of what you once said. Or do you change your mind from day to day, as I am almost beginning to think? My view of the matter, and also that of the Reich, is different. It is clear and straightforward, and not half-hearted. Nowadays there is no shame in a mother having an illegitimate child. On the contrary, it is the greatest happiness of a German mother. Those who take a

different view have nothing in common with the present age. Heil Hitler!

Looked at from this point of view, the problem of SS human stock-breeding assumes a different dimension. Without necessarily fully accepting the proposition that there was wholesale breeding by carefully selected men and women, planned reproduction in fact seems to have taken place.

The term 'breeding' first appears in an official document on 16 August 1943, when Dr Ebner uses the term *Züchtungsziel* (*Züchtung* means 'breeding', and *Ziel*, 'target, aim, objective'). A 'race expert', SS Sturmbannführer Lang, having criticized the methods of Lebensborn officials towards reproducers of both sexes, Ebner had the following circular displayed in all the offices of the organization:

We must always bear in mind that the SS is an organization of selected men, and that we in the SS consider the racial question in accordance with the principle of selection. Hence we take a stricter view than the party or the rest of the German people. The Reichsführer SS said in one of his speeches: 'The SS is a National Socialist military Order of Nordic-minded men.' That does not mean that only Nordic men – and of course women – can belong to the SS Order, but men who are marked by the Nordic mentality or have been educated in it. The objective of our breeding must be to bring together people from whom Nordic-minded children are to be expected. For that reason we reject the eastern and western races,* because the characteristics of the eastern race in particular are opposed to those of the Nordic race and because the mixture of Nordic and eastern results in people who are inwardly torn and unstable. He who has inferiority feelings because he does not look purely Nordic shows by the presence of those feelings that the eastern element in him is stronger than the Nordic.

When SS Sturmbannführer Lang says that our judgements are based only on racial appearance – it is the racial questionnaire that is referred to here – he is not correct. In my view the racial questionnaire is an excellent means of judging the total personality. Photographs show us the external appearance, but the answers to the questions give such a clear picture of the individual's

*In Nazi terminology, 'eastern' meant Slavonic and 'western' French, Walloon and other non-Nordic western peoples.

mentality that they provide us with a unique means of judging the racial body and the racial soul... The fact that the judges are not race specialists, but men and women with a good knowledge of reality, certainly does not in any way impair the value of the questionnaire...

In my opinion there is no occasion to change our attitude on the racial question. However, I am in entire agreement with SS Sturmbannführer Lang when he says that all our colleagues of both sexes should be better trained in racial questions...

The 'SS children' – as they are still called today – were to be the advance-guard of the super-race that was to populate the Reich that was to last a thousand years. They were the result of an incessant propaganda campaign in favour of sexual promiscuity for the sole purpose of propagation in the name of the race. With few exceptions, the men and women involved were volunteers. Both wholeheartedly involved themselves in what the American judges at Nuremburg called 'spiritual degeneration, an institution conflicting with the morality of every civilized nation'.

After the defeat, the Lebensborn mothers, from having been the 'sacred cows' of the Nazi period, became the most anathematized women in Germany. Today they say that they cannot understand why they were so ferociously condemned by popular opinion, and wonder why the fact of having had a baby in an 'ordinary maternity home' is still regarded as the worst of crimes. The question why Lebensborns are connected in the popular mind with human stud-farms offends and angers them, and they all say: 'I was engaged and was going to be married, but he died in Russia.'

Things get more difficult if they are asked, not about the fiancé, whose death always occurred on the Eastern Front, and so cannot be checked, but about the child. At this point the inquirer is invited to mind his own business, unless the woman tries to persuade him that the baby was taken from her by force.

The most concrete evidence that seems to support the 'human stock-breeding' hypothesis is a letter addressed by Himmler to his aide-de-camp Brandt from the Führer's headquarters on 8 May 1942:

The following points are to be discussed in Munich . . . Point no. 4. . . . a big headquarters for the Lebensborn. I have secretly given SS Standartenführer Sollmann orders to plan and construct the headquarters on the assumption of the approximately 400 000 women who already at this stage will be unable to find husbands because of the war and its losses. The style of the building must be in accordance with the noble and honourable idea of the unmarried mother.

8. 'Fountains of Life'

'Sacred shall be every mother of good blood.'
From the articles of association of the Lebensborn Society

Of all the Nazi organizations, the Lebensborn is still the most mysterious and the most controversial. Those directly involved still insist that it was established 'to eradicate the idea of abortion among the people' and that it intervened only at the stage when its clients were already pregnant. Frau Maria, who was secretary of the Lebensborn register of births and deaths, said:

'The Lebensborns were said to be human stud-farms, where men and women were brought together to beget children. That is false. The mothers who came to us were all pregnant already . . . Babies were made exactly as they are today, without the intervention of any institution.'

The man in the street took a different view. He believed that, if a girl had no partner and met all the racial requirements in force, she could present a child to the Führer with the Lebensborn's aid.

Everything points to there having been three different backgrounds to the business of conception.

1. *Natural reproduction*

At the outset the clients were mostly girls or women some months pregnant or on the point of confinement. Fleeing paternal wrath, the reproaches of the pastor or the priest, or public shame or the fear of it, they applied to the Lebensborn either at Munich or directly to Ebner at Steinhöring. Most came from Bavaria for the simple reason that that was where the first home was established. The system of racial investigation then began, not only of the girl but also of the man whom she designated as the father, unless the latter was a member of the SS or the police. One in every three applicants for admission was turned away; when the specifications for selection became stricter, one in every two applicants was rejected. If the prospective mother passed this hurdle, it meant that her child would, in the race experts' view, be of pure Nordic blood, of the same racial value as its parents, and she could await confinement in the best conditions.

This source of applicants increased proportionately to the results attained from the conditioning of women by the Nazi propaganda machine. It spread all over Europe, drying up only in 1945.

2. *Planned Reproduction*

Some girls, impelled by the continual propaganda to yield to their maternal instinct, applied directly to the organization to find out 'what they must do to have a child'. They had heard friends say that the Lebensborn was the agency chosen by Adolf Hitler himself for putting into practice what was known as 'service to the Führer'. The way in which the Lebensborn headquarters dealt with applications of this nature can be regarded as positive evidence of the existence of Himmler's planned reproduction programme.

All the witnesses' statements we gathered point to the recruiting of candidates for maternity having been carried out in the most innocuous-seeming manner possible. Analysis of the voluminous correspondence devoted to the recruiting of 'student nurses' in the light of the witnesses' statements (to which we shall refer shortly) makes it possible to dismantle the mechanism and, above all, discover the key-word in the Nazi terminology in the matter. For, just as 'disinfection' meant euthanasia and 'resettlement' extermination, so 'nurse' (*Schwester*) was used to indicate the voluntary guinea-pigs of the positive side of the population policy – the creation of the super-race. Everything that promoted that policy was as highly secret as it was possible to make it, just as was its negative counterpart, the liquidation of inferior races. Both were discussed only orally.

The inevitably blonde candidates for the post of 'pupil

nurse' were interviewed in the first place by 'reproduction advisers' at the office in the Herzog-Max-Strasse. These had to decide whether the girls' physical appearance was such that they would be likely to come up to the racial standards required of mothers of 'good blood'. If the decision was favourable, they would be asked to present themselves for another interview, armed with evidence of their Aryan blood. Their pedigree had to be traced back to the Thirty Years War, their parents had to be in good health, and there must be no Jewish blood in their veins. Finally the day of their physical and psychological examination arrived. Their blood, parents, ancestors, ideals, loyalty and devotion to the Führer were inquired into, and nothing, not even God, was left to chance. Many were so fanaticized that they signed a declaration abjuring Christianity in favour of the new religion of blood.

Documentary evidence of these abjurations is not available, but traces of them can be found in the correspondence between the girls and the Munich head office. The documents were filed at Steinhöring II, the Lebensborn's own register office. The Steinhöring establishment, of which Dr Ebner was in charge, was no ordinary maternity clinic, but the point of departure of all the experiments in the field of reproduction that were carried out by the Lebensborns. It was also at Steinhöring that the percentage of children left to the organization by their mothers was highest, though in principle Himmler preferred that a girl should keep her baby instead of 'irresponsibly . . . quickly getting rid of it . . .'.

The future of a 'pupil nurse' who surmounted all the hurdles and secured employment in a home was settled by a decision of the RuSHA dated 1937. The Lebensborn would bring her into contact with a man, 'not for any irresponsible purpose, but in full awareness of fulfilling a lofty duty to the nation'. She might meet him at the home, where doctors – SS doctors whom Himmler would not release for the front – sometimes outnumbered patients, or at a neighbouring SS convalescent home. There was no lack of opportunities, provided she abided by the directives – orally communicated – of the 'reproduction counsellors'.

How many girls responded to the oblique encouragement of that worthy *paterfamilias* Dr Ebner, who was the veritable linchpin of the whole edifice? At Steinhöring alone he supervised the birth of more than three thousand children. He never divulged any information on the subject, but according to Frau Ebner, 'many, many children' were born there between 1936 and 1945.

According to statistics compiled after the war, which documents currently available neither confirm nor refute, several hundred thousand women or girls were involved with the organization in one way or another. The figure of 12000 births, which it is impossible to check but which is sometimes accepted by historians who agree on the 'charitable organization' hypothesis, was put forward, without supporting evidence or documentation, by Sollmann at the American trial in Nuremburg of members of the RuSHA and the Lebensborn organization in 1947–8.

However, a number of clues enable us to draw a more accurate picture. On 22 January 1940 Ebner wrote to Sturmbannführer Pflaum: 'At the moment our homes are so overcrowded that our needs are becoming very urgent.' The Wienerwald home in Austria had 192 mothers on the roll in 1941, 250 in 1942, and seventy-two in the first quarter of 1943. Brandt, replying to a letter from the chief of police in the Poznan area asking for information, wrote on 30 April 1943:

The following Lebensborn homes are now active as maternity homes: the Harz home at Wernigerode; the Hochland home at Steinhöring, near Munich; the Kurmark home at Klosterheide, near Berlin; the Pommern home near Bad Polzin; the Schwarzwald home at Nordrach (Baden); the Wienerwald home near Vienna . . . All these homes are at present overcrowded . . . In the course of this summer it should be possible to open the biggest Lebensborn maternity home, the Taunus home at Wiesbaden . . . As for the other establishments throughout the Reich and Europe which are strictly reserved for the SS, they have no shortage of work.

Since many of these homes scattered about the country cannot now be traced, it may be assumed that some ordinary

SS hospitals had Lebensborn departments or annexes.

Three registers at Steinhöring recording the births of more than 2000 children miraculously escaped destruction. They were found by the Americans and later handed over to the new burgomaster they appointed. These registers are highly instructive, as they give the name and rank of the 'unknown' father as well as full details about the mother; they also enabled the present village clerk, Herr Karl Stabernak, to make an important discovery. According to these documents, the great majority of the women who gave birth to illegitimate babies at Steinhöring between 1936 and 1945 came from Munich, and, even more surprisingly, most of them gave one of the four following addresses: Kurfürstenplatz 1; Adelheidstrasse 26; Boschetsriederstrasse 10; and Ismaningerstrasse 95.

Unlike a number of other places mentioned in the files the function of which has never been explained, these four do not appear in any letter, circular or other document concerning the Lebensborns, but only in the records kept by SS officials in the Steinhöring register office.

What lay behind this new mystery? Was it just one more elaborate precaution taken to keep secret the identity of unmarried mothers of good race? Or were these simply accommodation addresses to simplify the girls' access to the homes?

A call at 26 Adelheidstrasse early in July 1973 drew a blank. As we found out, the tenants of the friendly-looking bourgeois house had lived there during the war but they did not even give us a chance to explain the object of our visit, which was clearly unwelcome.

However, the owner of the lodging-house at 5 Kurfürstenplatz distinctly remembered an SS woman tenant who had occupied a big, elegant flat at 1 Kurfürstenplatz until the end of 1943. 'She was a large, very severe and important-looking woman. She worked for the Lebensborn organization, and her name, if I remember correctly, was Viermetz, Inge Viermetz. I still have my books from those years, I can easily trace her, if you like.'

Now, Inge Viermetz had been in charge of the establishment of new homes throughout Occupied Europe and later of the adoption of children kidnapped in the Eastern territories. We had already interviewed her in March 1973, when she had insisted on the charitable aspects of the work of the Lebensborn. Her presence here suggested that the Nordic young ladies from Germany and the whole of Western Europe were taken care of by the organization before their arrival in Munich. But did they go to Inge Viermetz's flat, or to 26 Adelheidstrasse, while waiting to go to the home? Were they already pregnant, or did they become so only after staying at these addresses?

Living at 10 Boschetsriederstrasse was Frau Tietgen, a blonde of about fifty with steely grey-blue eyes. She is the widow of the former head of the Lebensborn homes in Norway, and she was obviously anxious to nip any conversation in the bud. She told us that her husband had been murdered by the French in 1945 in a prisoners' camp at Mutzig in Alsace, information she claimed to have had from an SS prisoner at Mutzig who had served in the same regiment as her husband. Our brief but painful interview yielded no new information, unless it could be regarded as tending to confirm the theory that these were accommodation addresses meant to conceal the trail. We next went to the Bogenhausen district of Munich, to 95 Ismaningerstrasse, which lies in the midst of more than two acres of green parkland, protected from the sun and inquisitive eyes by a six-foot wall surmounted by a row of trees now more than a hundred years old. The house was an impressive one, built in the style of the robust patrician dwellings of the end of the nineteenth century. Since 1945 it has belonged to the *Land* of Bavaria, which uses it as offices for its finance department. There is a plan to demolish it and build high-rise blocks on the site.

The search for eye-witnesses in this district, which had suffered severely during the war, continued to be difficult. Most of the residents, new or old, shook their heads and said: 'Lebensborn? Never heard of it.'

At 93 Ismaningerstrasse, however, a petrol station conceals the entrance to a courtyard that leads to a modest but comfortable house, the upper storeys of which have the advantage of overlooking the park and no. 95. Hans Schafstaller and his wife Franziska have lived there since 1928. In spite of the trees, 'which had not grown so tall then', they have a splendid view of the 'villa'. On 24 July 1973, this charming couple were willing to revive their memories, which we reproduce as accurately as possible:

'During the war our neighbour was the Lebensborn, if you have ever heard of it. We didn't realize it at first, of course. A word too much or an indiscreet glance could have nasty consequences at that time. But gradually tongues loosened, and one day the caretaker, whom we had known for ages, gave us to understand that the most incredible things were going on there.

'In spite of the constant fear, we were naturally curious, and every now and then risked a cautious look at the place. It was then that we saw the girls. A great many girls arrived there. They were all big and blonde, very Nordic, real Aryans, as the term was at the time. Those ladies were not our kind, they were not like our Bavarian women; in our part of the world women tend to be short and dark. They certainly didn't come from these parts. Many arrived on foot, others in huge Mercedes cars with curtained windows that drove at speed straight up the main avenue. Then they shut the gates again, and there was nothing more to be heard. SS men with big police dogs guarded the place day and night, inside the park and outside on the pavement. The local people got into the habit of walking on the other side of the street, to avoid having to pass the SS guards. We were German too, of course, but we were always afraid of those Lebensborn people . . .

'Later on in the morning, just before lunch-time, we would see the girls chatting to the soldiers who lived there permanently. You could tell from the uniforms that they were senior SS officers . . .

'It took us a long time to put two and two together, but in the end we realized what was going on. The house was a kind of meeting place, an SS officers' club. Then we found out – how it's difficult to say after all this time – that the girls came from all over the place, but were sent here by the Lebensborn head office in the Herzog-Max-Strasse, in the town centre, where the cinemas are. People said it was there that they were picked.

'Whenever new ones arrived people used to say: "Look, there are some more new cows for our stud-bulls." I must say they were a lot of fine, strong young men. One could well imagine that a girl who got a bit tipsy, let us say, might well want to give a child to the Führer, as the saying was at the time, with those young men. Because that's what the Lebensborn did. It helped girls who wanted to become mothers "for the sake of the Fatherland" to find an ideal partner – ideal, from the racial and political point of view. You see what I mean? There was a popular saying at the time that if you wanted a child with fair hair and blue eyes, all you needed to do was to apply to the Lebensborn . . .

'So far as one could tell from their clothes and other details, the girls used to stay for several days, sometimes for several weeks. Then they vanished, and new ones arrived. The men changed too, but not as often. It was known here that the girls went to Steinhöring, which is about thirty kilometres away, and there must have been an SS maternity home there. But we never found out what happened to the men after they left here . . . What did Hitler want with all those children? New regiments, no doubt about it. Throughout all those years we were afraid, we were simply terrified at having such neighbours. The Lebensborn people would have made short shrift of anyone who discovered their secrets. As you know, Dachau is not far away. . . .

'And as for Dachau and the precautions the SS took to keep everyone out, I can tell you something that will make you realize the cruelty of those people. Concentration-camp inmates worked in the garden here, and also on the heating. But they were never allowed in the house. And then, when the terrible air raids of 1943 and 1944 took place, the SS and their girls went down to the basement – which incidentally still exists today – and took cover there, while the concentration-camp inmates had to stay outside in the park. The result was that many of them were killed by the bombs. Afterwards they took the bodies away, heaven knows where, in the intervals between raids. But every time there was a raid it was the same . . . So now you know what the Lebensborn meant to us.'

3. *Artificial Insemination*

At this point a sinister character who played a considerable part in the programme for increasing and multiplying the German race makes a brief appearance: SS Gruppenführer Dr Conti, who also enjoyed the title of Reichsgesundheitsführer, literally 'Reich Health Leader', or Health Minister of the Third Reich. He committed suicide after its collapse. Eminent race theorist that he was, in the present context he interests us only to the extent that he became the régime's specialist in increasing female fecundity. His wife, Nana Conti, was president of the Nazi midwives' association.

Like all those who took an active part in pushing the policy of rapid population growth, Dr Conti had ideas of his own. Soon after the beginning of the war he proposed countering the 'small number of children in German families' by establishing matrimonial agencies under party control to prevent people from choosing their own marriage partners. Leaving this vital matter to individual choice was, in his opinion, detrimental to the will to reproduce. He said in a speech that Germany in the future would have colonies or distant territories in which women would be in short supply, so that men would have difficulty in finding suitable marriage partners. War losses theoretically did not necessarily lead to a diminution of numbers in future generations, since the birth rate depended exclusively on full exploitation of women's fertility. After giving much thought to the problem of enabling unmarried racially valuable women to have children, he had decided that there was a solution that could be 'recognized as honourable by the people and the State'. Like other master-minds of Germanic ideology before him, he had considered various alternatives before (being a medical man) coming down heavily in favour of exploiting a medical discovery which, in his opinion as a doctor, would change the whole nature of the problem. This was artificial insemination, which the Russians and Americans had been using in cattle breeding for some time.

He became a most ardent advocate of this 'revolutionary method'. Though he knew Himmler was against it, he believed he would be able to overcome the Reichsführer's scruples. Its chief advantage, in the eyes of this alarmingly naïve Health Minister, was that it did away with 'the psyschological complex of the sexual experience'. He continued:

> True, as a result there would be a rather soulless, mechanical element in the process of fertilization, but in the last resort it is at least just as unnatural for healthy, vigorous women not to be able to indulge their natural maternal instincts and have to remain childless.

Conti was not the only personality of the régime to toy with the idea of this new method. According to Berthus Brandes, a colleague of Kaltenbrunner's, who gave evidence at Nuremberg, the possibility of using artificial insemination to accelerate the production of Nordic Germans was discussed at a Führer's conference held in Berlin on 16 July 1941 to discuss the population problem. A letter of March 1943 to a Professor Henseler gives a general idea of Himmler's attitude:

> Like you, I take the view that we in Germany must not be left behind in any field or any area of research. Nor must we be left behind in any matter, even though our attitude to it is one of total rejection. In the field of artificial insemination in particular, there are some things we must be perfectly clear about. There can be artificial things only where we, by reason of our culture, civilization, domestication and the sum-total of our intervention in nature, wish to participate in nature's own work. I believe that artificial insemination should be used in the broadest way for all kinds of animals that are destined for the slaughter-house or serve similar purposes . . .
>
> But I believe that artificial insemination possesses to a high degree all the disadvantages that you enumerate, above all the impoverishment of hereditary or genetic factors. What speaks against it in my view is the belief that we human beings cannot

improve on nature. If nature in its wisdom decided to produce that vast mass of germinal cells and ruled that the whole sex act, with all its physical and psychological peculiarities and characteristics, was necessary for reproduction, I think that artificial fertilization, by dividing the germinal cells into twenty or a hundred parts and smearing the uterus of twenty or a hundred female animals with them, is perfectly practicable. Nevertheless I am firmly convinced that it is bound sooner or later to lead to the deterioration of future generations, and probably to impotence or sterility. . .

I approve of it in the production of animals that in general do not have to pass on their hereditary factors. It can be an emergency solution in special cases when no male animal is available. But I regard it as dangerous for our breeding and the handing on of our hereditary factors as a whole. In buying a domestic pet of any kind I myself always prefer a product of natural and not artificial insemination.

This also applies to the field of human reproduction. The laws of all the Indo-Germans show that they were aware of the problems of male sterility. They were certainly advanced enough in knowledge to be aware that fertilization was possible by artificially introducing semen into the female body. But, because of their high ethical level, they anchored the natural processes of reproduction into their basic moral and legal code . . .

These are only a few thoughts on the great problem which, as we know, is regarded very differently in America. Heil Hitler!

On the other hand, on 4 November 1944 Dr Ebner replied as follows to Frau Elisabeth H. in Berlin:

I have read with interest your letter in which you say you have decided to have another child, but would like to choose the method of artificial insemination for the purpose. You suspect that your husband whom you love is not capable of giving you one. As you find it hard to decide to have relations with another man, the first step should be a medical examination of your husband's semen. Should it turn out to be as you suggest, why choose a mechanical and purely impersonal way of having a child? You say you have already decided to choose a man other than your beloved husband to have a child by, but that at the last moment you have inhibitions, because you cannot be 'frivolous'.

At this point I must ask you a question. Is it really frivolous if a woman deliberately goes with a man in order to have a child? I believe that on the contrary, it is acting with a deep sense of responsibility, and that such a woman shows a great deal of personal courage and firmness of character.

In your circle of friends and acquaintances you certainly know a man with whom you could calmly and frankly discuss this matter.

This woman was a former Lebensborn client ('you say you have decided to have *another* child') who must have regarded artificial insemination with the aid of the Lebensborn organization as a practical possibility. Otherwise why write to Ebner in such obvious terms? Remember that Ebner's letter was written in November 1944, when Germany was reeling under a hail of bombs. Railway and telephone lines were cut, communications were difficult, and Ebner could no longer travel freely between his 'beloved homes', some of which had had to be evacuated from occupied countries and brought back to a Germany whose *Lebensraum* was rapidly shrinking. He was no longer able to give 'oral' advice to his clients, and so he did so by mail, but weighing his words as if he already feared they might one day be read by indiscreet eyes. Hence an outspokenness not usually to be found in his correspondence. This faithful servant and friend of Himmler's believed that one extra German child might yet help to turn the scales in Germany's favour. Embarrassed, for 'there are things it is difficult to write about', but horrified at the idea of Frau H.'s not having another child, he tried to teleguide her back to the natural method of reproduction.

Right up to the collapse of the Reich, rumours circulated at all levels of the population about ways and means of combating the sterility by which some citizens felt themselves threatened. Because of their reputation for human stock-breeding, it was believed that only the SS had these ways and means available; and those who needed them did not hesitate to apply directly to the Lebensborns, as is shown by Dr Ebner's correspondence, or to an SS officer they believed to be in a position to intervene effectively on their behalf.

Here is a letter, dated 12 November 1944, from the head

of the gynaecological department of a hospital in Schleswig-Holstein, addressed to Herr Fotograph Heinrich Carstensen of Rendsburg:

May I be permitted for once to ask for your advice in your capacity as an SS leader? The provincial gynaecological department of which I am now in charge has been given the special task of providing assistance in childless marriages. There have been an increasing number of cases recently in which young couples still want children after the husband has become sterile because of some unlucky chance, and they want the child to be at least the wife's. Thus, for instance, Frau B. writes to me as follows: 'Is there any such possibility for us? Is there an institution in the Reich where one could beget a child with a healthy and uninvolved man?' Has the SS such an institution, or could you inquire of your superiors, or give me an address to which I could write?

It is of course difficult to decide whether one should advise the woman to resort to artificial insemination or advise, for instance, the course suggested by this couple. Can you give me any advice in the matter? Heil Hitler!

Thus this gynaecologist seems to take artificial insemination for granted, though he also seems aware of the possibility of making arrangements through the SS by which his patient might be able to conceive a child 'naturally'. Unfortunately, no reply to this letter has survived in the files.

9. 'The Little Blonde Sisters'

> 'Oh, how proudly did these adolescent girls push their loaded, feather-bedded prams along the city pavements, showing off the living evidence of their "service to the Führer and the nation" and challenging their contemporaries to similar immorality and surrender of their virginity.'
>
> CARDINAL GRÖBER, pastoral letter of 1 August 1945

Though it was a veritable state within the state, with complete financial and administrative autonomy, its own register office and its own ideological education of mothers and children, the Lebensborn organization, surprisingly enough, did not enjoy complete freedom of movement.

On the level of power relationships, there was a strange parallel between Nazi Germany and the Germany of the old days. The petty dictatorial kingdoms and minute duchies whose squabbles and wars prevented the geographical and political unification of the country until 1870 reappeared in different forms and squabbled for power or Adolf Hitler's favour. The party would be at loggerheads with the Wehrmacht, the Wehrmacht with the SS, the SS with the Ministries, the Ministries with the Abwehr and the Abwehr with the SD, unless the Ministries were fighting a private war with the Wehrmacht or the party was quarrelling with the Abwehr; all of them were greedy for power.

Even the creation of the super-race, at any rate in the form envisaged by Himmler, was not always unanimously supported, and the Lebensborn Society encountered obstacles. As the contagion spread, often among very young girls whose lives were seriously affected, public reservations also grew; in particular, hostility was roused by these 'ladies of luxury', whose life-style was so different from that of other German mothers. In this hostile atmosphere it became by no means easy to recruit qualified staff.

The files contain categorical refusals by Red Cross nurses to work in the Lebensborns and – even more surprising – similar refusals by members of the NSV, the Nazi welfare association (though we shall deal later with the NSV 'Brown Sisters' who took an active part in kidnapping Polish children). But for the bad reputation of the Lebensborns, Ebner's and Sollmann's frequent cries of alarm about the shortage of qualified staff, their numerous requests for aid addressed to Nazi higher institutions, would be incomprehensible.

Nazi organizations such as the German Red Cross or the NSV did not question Himmler's race policy in principle, but had difficulty in swallowing its methods. The creation of the super-race required the use of German guinea-pigs, which the older generation did not find easy to accept. Ebner, evidently aware of this, took the precaution of taking with him to Steinhöring an elderly midwife who 'gave him loyal service for fifteen years'; elderly and long since retired midwives agreed to go back to work in the Lebensborns only after long hesitation.

The following circular was sent by Dr Ebner to all the maternity homes on 2 September 1941:

> A few days ago a nurse in a Lebensborn home had to be instantly dismissed for striking a ten-month-old baby in the face so violently that the bruises were visible for days afterwards. This incident prompts me to ask heads of homes and head nurses again emphatically to remind all nurses and nursing staff that chastisement of children is forbidden in the Lebensborns. A nurse puts a black mark against her qualifications as a bringer-up of children if she cannot manage a baby under the age of one except by brutally striking it. There are recurrent complaints that children who are slow eaters have their noses held to force them to swallow. This method too comes under the heading of maltreatment of children and must be strictly forbidden.

Thus these maternity homes, unusual in themselves, employed 'nurses' whose conduct and professional skills were also out of the ordinary.

The 'nurses' for whose benefit this order and many others were issued were not professionals, or probationers, or even pupils. They were merely girls who were employed by the Lebensborn before or after giving a child to the Führer as some sort of material reward for their services. There is no doubt about this. The word *Schwester* in German means 'sister', 'nurse' or 'nun' (member of a religious order). This ambiguity enabled the Lebensborn organization to use it comprehensively for nurses, untrained ward assistants and women members of the 'fraternity of blood', meaning women members of the Black Order, the SS.

Only one of these 'nurses' whom we were able to trace in Germany thirty years later was still exercising her profession. All the rest had jobs not even remotely connected with medicine: they were cooks in restaurants or school caretakers; one was caretaker at a Jewish community centre. Most were living a normal family life with several legitimate children. An official report dated 1 March 1941 stated: 'After the war all our nurses will have to serve their apprenticeship and not know only how to give injections. At present we should be glad if 50 per cent of our nurses were professionals.'

As for the doctors at these institutions, doubts exist about the qualifications of at least some of them. Nurses' complaints about them were numerous, and these two examples are typical:

Dr Laumann is a magnificent organizer, a brilliant talker and a real comedian, but as a doctor he is practically non-existent.

The patients boycott our obstetrician, because he has only a dentist's diploma.

The fact is that the nurses were not judged by their professional skill. In September 1938, Sister Ida T. was arbitrarily dismissed from the Wernigerode home. The correspondence shows that she protested, but she had to go. She had merely failed to come up to the strict racial standards required of the Lebensborn staff. Next day an order was put up in all offices saying: 'Henceforward only 100 per cent racially valuable sisters may be employed.'

The incomplete correspondence does not reveal whether the race expert who examined Ida was reprimanded for the unpardonable crime of passing her. All we know is that she 'had no child'. So perhaps the gynaecologist who did not discover her sterility was at fault. Poor Ida, being sterile and not the Nordic type, had no chance.

A contrasting case was that of Adelheid K., who was sent to the Ardennes home, near Liège in Belgium, in the autumn of 1943. She was described as dirty and untidy, and after a few weeks was dismissed for 'having done nothing to check an epidemic of diphtheria that had already cost the death of a child', and a German child at that, and was declared unsuitable for work in a Lebensborn. The child was buried at Liège with military honours. One might have expected this to have been the end of Adelheid's 'nursing' career. Far from it. She was transferred to a home in East Prussia and then to a home for kidnapped children at Kalisz in Poland. She owed this to Dr Ebner, with whom she conducted an extensive correspondence. She had presented one child to the Führer and was preparing to present him with another.

As most of these 'nurses' entered the Lebensborns as expectant mothers of illegitimate children, it explains why the word *Schwester* and the initials KM (for *Kindesmutter*, 'mother of a child') so often appear together. That they changed homes a few weeks after their confinement was presumably to obliterate the trail which might have led 'enemies of the Reich' – such as their parents – to discovering their whereabouts.

These Lebensborn girls were blonde, were good child-bearers, had passed the tests of the selection committee and had to belong to the élite and be at least 1·60 metres in height. The only thing absolutely forbidden them was sexual relations with men unknown to the race office. When they became mothers, in the Lebensborns or elsewhere, sometimes ten, twelve or fifteen months after the beginning of their employment, and as a result of the attentions of a doctor, male nurse or visitor belonging to the SS or the police, they were assured of the best possible care. But woe to those who responded to Hitler's appeal without taking the necessary precautions, in other words without consulting the 'reproduction counsellors' first.

One day a 'nurse' turned up at Steinhöring in the hope of having her confinement there. She told the interrogating officials that she had spent a night with a soldier in Vienna. Next morning he had gone out to buy her a present and had never come back. She remembered the name of the hotel, and the Gestapo made inquiries there immediately, but the only address the soldier had left was his field post number. Was he of pure Aryan blood? That was the only thing that interested the inquirers. Weeks passed without a trace of the seducer. The desperate girl had to appear before a RuSHA committee and declare under oath that she had not had relations with any other man at that time. Being blonde and racially valuable, she was eventually appointed a nurse in a home, 'pending her confinement'. But the Lebensborn refused to sponsor the child, because the father was unknown.

Let us linger for a moment over this example of a refusal to sponsor the child of an unknown father. Isolated from the general context, it might seem a minor detail, yet it has a strange aspect. The documents show that of every ten children born in a Lebensborn, eight were registered as of 'father unknown'. We have been able to find no other instance of an inquiry of the same kind. Certainly there are instances in which paternity was checked by the authorities or denied by men who wanted to avoid paying the maintenance contribution. And complications of all sorts were caused by men who dropped the girl 'immediately afterwards'. None of these cases, however, was submitted to the Gestapo. Therefore the fathers of the children born 'in the Führer's service' cannot have been so unknown as was stated on the birth certificates that the Lebensborn itself filled in. The 'race experts' were not to be trifled with, and the best way of ensuring racial purity was to supervise and control reproduction.

So the racially valuable 'Little Blonde Sisters', whose nursing skills hardly extended as far as giving injections, entered the Black Order like candidates entering a religious order. The latter devote themselves to God and the former devoted themselves to the Führer, the god of the Greater German Reich, or to the Reichsführer, his brother; and several persons we interviewed actually spoke of Hitler as

God. Frau S., for instance, who presented a son to the Führer at Steinhöring in 1943, is now an active Jehovah's Witness. 'Then Hitler was my god,' she told us. 'Since then I have found the only true God.'

Such language owes everything to the ethos of the Third Reich, and of the Lebensborn organization in particular. Here is a letter from SS Standartenführer Dr Brandt to Frau Else Köhne at Berlin-Charlottenburg, dated 13 November 1944:

As a long-standing subscriber you sent to the West Berlin editorial office a description of a small experience with a view to possible publication. The Reichsführer SS has been informed of the charming story by the *Schwarze Korps*. He enjoyed it greatly. There is no question of its publication, as you will understand, but the Reichsführer SS would like to do something for the little girl that will give her pleasure later, so that she will look back gladly to the stirring time – which at her age hardly affects her – of Germany's great struggle for existence. A savings bank book is to be opened for her. Her childish mistake in believing that the Reichsführer SS is the brother of God can be corrected when the savings book is handed over to her when she is older . . .

If Germany had won the war, the Lebensborn would certainly have undertaken the establishment of innumerable villages for unmarried mothers. At the Lebensborn trial at Nuremberg Sollmann said that, by Himmler's orders, the organization was to have built 500 wooden houses at Smoscewo in Poland for exclusive use as mothers' and children's homes. He said that the order had been given in 1939, and that work on the project had in fact begun. The site was actually marked out on the ground, but only a few houses were built. The court was given no evidence about the project apart from Sollmann's statement. When we mentioned it to Paula Hessler, who was a genuine nurse in charge of babies, first at Bad Polzin and later at Steinhöring, she commented:

'The Lebensborns were Himmler's hobby, you see. He had quite fantastic plans for them for the future. His first concern was for the children, but he also thought of the mothers; he wanted to enable them to live with their children. He wanted to establish homes for them. Some already existed, there was one at Stettin, for instance. I heard a great deal about those places, where nurses looked after the children while the mothers were out at work.'

Himmler also had 'fantastic plans' for his SS women auxiliaries; he wanted to be the founder of an entirely original institution, halfway between barracks and brothel. Meanwhile he instructed his carefully selected young ladies to devote their spare time to sporting and cultural activities. They were not in the habit of disobeying orders, with the result that their cultural exchanges, particularly with Luftwaffe pilots or wounded SS men in hospital in the back areas, in many cases gave them the opportunity temporarily to give up wearing uniforms that in any event were getting too tight for them. If they were then posted for a short while to a Lebensborn home, they and their Luftwaffe counterparts were able to continue their cultural activities. Every evening they virtuously attended special courses and lectures for mothers, imbibing the doctrine of human stock-breeding from such masters of racial theory as Goebbels, Conti, Darré and Himmler.

On 13 June 1940 a party of fifty pupils from the training school for nurses at Tutzing paid an instructional visit to the Lebensborn at Steinhöring. After being taken round and shown everything in great detail, they were treated in the afternoon to a long lecture about the work, aims and national utility of the Lebensborn Society. So enthusiastic were the girls about what they had heard and seen that, after their return to Tutzing that same evening, they organized a discussion, and one of them was appointed to draw up notes to be presented to the school authorities. The following are selected extracts:

> Lebensborn – an important problem for the country. Views still divided. Things that rouse opposition not necessarily bad, e.g., the [Party] time of struggle. What counts is results.
>
> No difference between married and unmarried women when both are mothers.

Extramarital relations are tolerated so long as the girl is not pregnant. Phoney morality.

But pregnancy is the natural consequence of a love relationship.

The great blood-letting of our people through the war – replacing the losses – results more important than legality.

Annual births 1·2 million, number of abortions half a million – (enough to fill the Olympic Stadium).

Women who are mothers of children frequently get married.

A woman's chances of marriage not primarily dependent on whether or not she is a mother, but a question of supply and demand.

Ideal state of affairs – one man for one woman.

Two million surplus women. Still more because of present war.

Shortage of men accentuated by homosexuality; a million of them.

Thus three million men too few. Rape of the Sabine women.

Every man has about five women to choose from. Presumably he will choose the prettiest and richest who is also childless. Should the other four remain childless because they have no hope of marriage?

The problem goes much deeper than that – is our people to be or not to be? Germany must again become a land of children. The unmarried girl forced to remain unmarried must have the right to motherhood – hence our fight against defamation.

Is our struggle justified? Answer: Good is what is useful to the nation. Bad is what is harmful to the nation. The nation is not just we the living, but our forefathers and also our descendants.

Every mother of good blood must be sacred to us.

Many of the girls who took part so eagerly in this discussion were under sixteen.

10. The Reichsführer's Tears

'The child Sigrid M. creates the impression of complete idiocy. Adoption would be difficult. It would be better to hand it over to the mother for bringing up.

'Little Horst W. is an imbecile. At the age of two he does not talk yet and grabs other children's toys or sits on his stool and waggles his head. He thinks only of eating. What is to be done with him?' (No answer in the files)

To the reader of these documents the most striking thing is not the arrogance of these mothers, or the primitive methods of the dentist-obstetricians, but the huge correspondence between the various Lebensborn departments that followed the illness or, in particular, the death of a single SS baby. On 7 July 1940 Himmler signed a circular requiring a thorough investigation of every case of death in a home. In spite of the Reichsführer's orders, his repeated visits with armfuls of presents, the care lavished on the girl mothers, the quarantines, the forbidding of visitors and the bans on leaving the establishment, the death-rate among babies of superior blood in the Lebensborn clinics was high. It seems, in fact, to have been about 8 per cent in comparison with the German average at the time of 6 per cent for the same period. Many death certificates, all signed by Ebner, are still in the files. For every baby boy carried off by meningitis or every baby girl who failed to survive pneumonia, there are ten, twenty, thirty or sometimes more letters explaining how, why and in what circumstances death dared to strike a representative of the super-race. The investigations carried out by the 'race experts', the post-mortems (insisted on by Himmler, who flew into a rage if he were not immediately informed of a child's death), Ebner's, Sollmann's and Himmler's letters of condolence, the discussions that took place about burial costs or the efforts to find a 'vanished' mother, inspire an insuperable sense of revulsion when we recall how millions of other children in Europe died.

From one file we learn that the death of a Lebensborn child caused the Reichsführer's SS's eyes to fill with tears. The reports on births and deaths drawn up for his benefit were continually falsified to 'please' him. Though he insisted on being informed of every case of death or serious illness, he wanted to hear nothing of abnormal children who did no credit to his labours; he specifically forbade such reports to be submitted. Physically or mentally handicapped children who were born in the Lebensborns, or those who contracted rickets because of inadequate care by irresponsible 'nurses', were the bane of the organization.

Wienerwald Maternity Home.

The child Siegfried B. appears to be in good health, but he often screams until he becomes blue in the face and then rolls his eyes in an alarming manner.

I must further inform you that the mother herself made a bad impression, and according to our SS principles she should not be allowed to have another child.

This child was transferred to a Vienna hospital, where it was found to be suffering from a squint, which was treated. A nurse named Ruth O. gave birth to an abnormal child in 1941 and had to give up her job, 'there being no place for the child in our establishment'. There were many such unfortunate children, born with a harelip, a goitre, or a club foot, or afflicted with deafness, water on the brain or mongolism. Even before their fate was settled, often brutally, the first thing that was done was to carry out a thorough inquiry into the mother's and father's parents and grandparents, as well as their brothers and sisters. If the father were a good Nazi, a senior SS officer, the 'L office' authorities made it their business to discover an accidental reason for the abnormality. In the last resort the question was submitted to Himmler, whose decision depended on the father's ideological value. If the 'ideological verdict' was negative, the child's parents were forbidden to have any more children.

At the Harz home at Wernigerode the wife of a concentration camp guard in Poland gave birth to her fifth child, which was abnormal and had an incurable illness. Mother and child were expelled immediately. Inquiry showed that the couple had had two other children who had died of the same disease, whereupon the father was dismissed from his job and struck off the roll of the SS.

But if a child 'of good blood' had a curable illness, herculean efforts were made on its behalf. Professors were put to work to find ways of curing club feet, operating on goitres and harelips and all the visible defects which were embarrassments to the eugenists and diminished the chances of adoption; and invariably there was a voluminous correspondence, full of the most detailed information, warnings to take the greatest possible care, desperate appeals to the miracle-working doctor to bring about a swift cure. Aryan blood was precious indeed in the Lebensborns. It was collected in test-tubes, sent from Munich to Berlin, from Berlin to Leipzig.

No such care was lavished on children of 'doubtful' origin. On 30 November 1943 Ebner replied as follows to a Leipzig doctor who had naïvely inquired whether the Lebensborn organization would undertake the adoption of a child born deaf and dumb:

> The rules of the Lebensborn are that only racially and biologically valuable and hereditarily unimpaired children can be accepted into our homes.

As we have seen, the progeny of the master race was not spared its share of physical and mental defects and illnesses, both grave and benign, which are the lot of all other children, without regard to the colour of eyes or skin. Nor were they spared the congenital abnormalities that sometimes appear only after a certain age. But what differed from the normal was the reaction these cases evoked. Adoptive parents reacted with indignation. They felt they had been cheated, palmed off with second-rate goods, and promptly returned the unfortunate child to change it for a healthy one. Dr Ebner wrote as follows on 4 August 1941 to SS Oberführer Langleist at Buchenwald concentration camp:

The Lebensborn organization has entrusted me in my capacity as chief of medical services with the task of replying to your letter in regard to your foster-child [i.e. at the preliminary stage to adoption] Wolfgang.

I regretfully gather from the correspondence that your hope of obtaining a healthy foster-child has not been fulfilled in the case of little Wolfgang, as the child has developed a nervous instability that makes it almost impossible to give it a normal upbringing. I can understand that your wife, who has done her utmost to care for this child, has ended by suffering physically and mentally under the strain. I have therefore suggested to the head of the Lebensborn, SS Standartenführer Sollmann, that the child should be fetched and placed elsewhere.

You also expressed the desire to be offered a healthy child for adoption in Wolfgang's place. Unfortunately this wish cannot be fulfilled immediately. We have in the Lebensborn several hundred applications from childless parents who want to adopt children, and their requests are met in turn. Unfortunately only a few mothers are making their children available for adoption at the present time; besides, I do not consider just any child available for adoption would be suitable for you. I therefore do not believe we shall be able to meet your wishes in the matter in the immediate future, and I would therefore ask you to be patient for a time. Heil Hitler!

Thus the care with which the adoption procedures were followed did not prevent some of the adoptive parents complaining at not getting from the Lebensborn the *Edelprodukt*, the top-quality goods with fair hair and blue eyes which they believed themselves entitled to expect from that institution. Some threatened to write directly to the Führer expressing their dissatisfaction; still others went even further in showing their lack of human feeling towards these unfortunate infants. An engineer from the Graz area roundly accused the Lebensborn of responsibility for the death of a child he had been sent. It had rickets and was ill-cared for; it died only a few days after its arrival. In view of the obviously bad condition in which it had reached him, he refused to pay the funeral expenses.

It was on Himmler's direct authority that the Lebensborn

organization permitted the systematic elimination of all the abnormal children who, according to the principles of selective eugenic reproduction, should never have been born. Thus dozens, perhaps hundreds, of babies born in the Lebensborns were sacrificed, like sub-humans, the insane and undesirable human stock in general, to the gods of the Nordic race.

No need was felt to exercise any tact or sympathy in informing the child's parents of the decision to 'disinfect' their offspring; they did not even have the right to the small packet of ashes that in other cases of 'euthanasia' was sent through the post. There was no need to explain to volunteers in the Führer's service that the abnormal child had died of a bad cold or an inflammation of the middle ear. They did not care how it died, for it rid them of a taint. The father, whose identity would be miraculously rediscovered on the occasion, would do everything in his power to shift as much blame as possible on to the mother and retreat into anonymity as fast as possible.

The Lebensborn authorities were even more strictly bound than those of other maternity homes in Germany to report to the Reich health authorities all cases of congenital idiocy, mongolism, microcephaly, paralysis of various kinds, and all sorts of other physical and mental defects with a view to protecting the 'genetic wealth of the German people'. Particulars were submitted to the Reich Committee for the Scientific Investigation of Grave Hereditary Illnesses and Defects. Generally the little victims remained in the home for only a short time after birth. Without awaiting the results of the investigation of their parents that invariably ensued, they were taken in an 'L office' vehicle to a former prison that had been converted into a psychiatric clinic at Brandenburg-Görden, near Potsdam – an institution which still exists and where a small commemorative tablet recalls what went on in it between 1939 and the last days of the war in 1945. Here they were personally received by Professor Hans Heinze.

The speciality of the place was the 'disinfection' of infants up to the age of three (subsequently it also accepted sixteen-

and eighteen-year-olds), and the number of the professor's and his associates' victims is unknown. How many were supplied by the Lebensborns is also unknown. The cold and laconic terms of the documents about them are in striking contrast with Ebner's or Sollmann's effusions about a harelip or a case of German measles:

The Lebensborn Society Division A Munich 1, 7 February 1942.

To the Chief Medical Superintendent of the Lebensborn Society, SS Oberführer Dr Ebner, Steinhöring, Upper Bavaria.

Re the child Jürgen Weise, Pommern Home.

Ref. your letter of 3 February 1942.

Oberführer,

In reply to your letter of 3 February I have to inform you that SS Standartenführer Sollmann has decided that the child shall be taken to the *Land* institution at Görden. I have informed the home accordingly. Heil Hitler!

(Signed) Viermetz Head of Division A

To the Head of the Pommern Home, 6 March 1942

SS Obersturmführer Dükker, 1, Bad Polzin, Pomerania.

Dear Comrade Dükker,

As the *Land* institution at Görden seems to have failed to inform you, I herewith notify you for your information that the child Jürgen Weise died there on 23 February 1942. Heil Hitler!

It is now known that the children were not killed off at once. Their life was prolonged with luminal or morphine injections so that if necessary it might be claimed that every possible effort had been made to save them. As the committee's purpose was the improvement of the race, the bodies – of the Nordic children in particular – were dissected for the purpose of 'scientific research into grave hereditary and constitutional illnesses'.

These documents, which miraculously escaped destruction, provide definite evidence of the close collaboration of the Nazi organizations in carrying out the racial policy of the Reich. Lost in a multitude of anodyne documents, these two letters did not figure in the files produced at the trial of the handful of Nazi doctors who had to answer for their crimes at Nuremberg. They, with a few others, come from the three complete files of Dr Ebner's correspondence.

But let us look more closely into the matter. The letter dated 7 February 1942 from Inge Viermetz, the head of A Division, was addressed to Ebner, who was then at Steinhöring. A month later, on 6 March, Ebner informed the head of the Pommern home, from where the child Jürgen Weise came, that the latter had died on 23 February. Now, in the file the two letters are in chronological order, that is, no letter was addressed to Ebner between them. In other words, no one had informed Ebner of the child's death. But Ebner, knowing the procedure well, simply put down 23 February as the date of death. It is difficult to believe that a child, or rather children, taken to Görden in an official Lebensborn car, survived the professor's treatment for so long. Such was the complicity of the two organizations, no subterfuges were necessary. Moreover, there was no risk of any parents turning up at Görden to find out what was happening to their mongol son or deaf-mute daughter. The SS man and the blonde 'nurse' to whom such a misfortune had occurred would take good care to do nothing that might aggravate their case. But the administrative rule was the same for everyone. About a fortnight must elapse between a child's arrival and its death, the death certificate having been written out well in advance, sometimes on the day of its arrival.

The following story is no more than a tailpiece to this chapter.

Her name was Nini and his Sigmund. She was a close personal friend of the Reichsführer's, and he a medical officer in the Luftwaffe who by Himmler's orders had been posted to Dachau to conduct experiments 'in the interests of German troops' on the reheating of human bodies exposed to extreme cold, resistance to low pressure, coagulation of the blood, and human resistance to cold and over-cooling in general. Dr Sigmund Rascher, who had the rank of captain in the Luftwaffe, used methods which the Luftwaffe did not exactly appreciate, which was why at the beginning of 1943 he was transferred, as a result of Nini's tireless efforts on his behalf, to the SS, in which he became an Obersturmbannführer, equivalent to lieutenant-colonel, and felt entitled to regard himself as a special protégé of Himmler's. But for the requirements of the RuSHA, the couple might have gone on living happily together, though Nini was fifteen years older than he. But Himmler told them they would not be permitted to marry unless they were able to prove their ability to produce a numerous brood of biologically pure and genetically untainted children.

During working hours Sigmund devoted himself to his scientific experiments. In particular, he studied the copulation of Jews in certain special conditions. Twenty or a hundred times he put a naked Jew and a naked Jewess together in a special hut in the concentration camp for a longer or shorter period. The male had been previously refrigerated until he lost consciousness and was placed on a bed in this condition. Sometimes he would be given the privilege of being placed between two naked females. The results surprised and disappointed Dr Rascher, for he found that males came round more quickly with one woman only; he made the remarkable discovery that after coitus warmth returned to the male body more quickly. Some of the guinea-pigs died, of course, and of the three hundred that were used nearly one hundred refused the female offered them. Himmler encouraged his protégé to continue these fascinating experiments, as is shown by this letter of 13 April 1942:

> Your last discoveries seem to me to be very interesting. You should continue them on further prisoners condemned to death . . . Should any of these chance to survive, they can be reprieved and sent to a concentration camp for life.

For her part, Nini did not cease invoking all the gods of the Nazi pantheon in the hope of patriotically becoming pregnant in spite of her forty-eight years. Time passed, experiments continued day and night, but with no success. As can be well imagined, Rascher vented his fury on the victims put at his disposal by Himmler, who in a letter to General Milch of the

Luftwaffe in November 1942 considered his 'experiments to be of capital importance'.

One day, when four prostitutes arrived from the Ravensbrück camp to revive refrigerated Jews with their animal warmth, his indignation knew no bounds, because he noticed that one of the girls was a German of pure Nordic type; his conscience revolted at the idea of forcing such a girl to lie with sub-humans of the camps. But his ill-humour was of short duration, because Nini, 'in the strictest privacy', at last gave birth to a child. This initial success was crowned in due course by the arrival of two more children. The delighted Himmler gladly granted the couple an SS honeymoon, and declared his willingness to be the children's godfather.

Nini's triumph did not fail to find its place among the *Wundertaten*, the miraculous achievements, reported in the SS journal, and RuSHA gynaecologists hastened to examine the formerly sterile woman in the hope of discovering a method of enabling other German women to bear children after the menopause. However, Himmler ordered discreet inquiries to be made about this Nini Diehls, who in her youth had burned the candle at both ends; and, unfortunately for the couple, it was discovered that there was no value in the methods of the 'institute for research into the biology of reproduction and the combating of sterility'. Nini was sterile (unless it were Sigmund), and the three children had been stolen. Himmler, who godfathered only *Edelprodukte*, goods of outstanding quality, felt he had been grossly deceived. Rascher was sent to the Bunker of the Dachau camp, where he was apparently liquidated in April 1945, and Nini was hanged at Ravensbrück.

11. By the Sword and by the Cradle

'It would be highly desirable for the breeding of our children to bring racially and politically valuable Norwegian women to Germany.'

Letter from Ebner to Himmler, September 1940, quoted at the Lebensborn trial

The aggressive policy of the Third Reich finally bore fruit in the *Blitzkrieg* conducted by a meticulously prepared Wehrmacht initially against the West. The five years from the Austrian *Anschluss* of 12 March 1938 to the defeat at Stalingrad in the spring of 1943 were marked by one German military triumph after another.

In the East Nazi ferocity in the service of its negative population policy was exercised without restraint. Eleven million Jews and thirty million Slavs, to whom several million 'incorrigible' political detainees could be added, were due for extermination by 1980. Seventy million other sub-humans were to be transferred to Siberia, Africa or Latin America.

The object of the positive population policy was to replace these 'disinfected, resettled or evacuated' populations with members of the master race. By breeding or 'recovering' a hundred million Nordics, it would be possible to attain the total of 250 million Aryan Germans aimed at by Himmler and his experts, also by 1980. Moreover Himmler, while addressing SS and police chiefs at a lunch on 16 September 1942 on his population policies, spoke of the repatriation of people of German stock from America and Africa after the war. In the course of the century there would be from 400 to 500 million people of German stock to populate the *Lebensraum* that Germany would have acquired.

In 1940, after the campaign in the West was over, the Reichsführer, thinking of the Nordics in Norway, Denmark, Holland, Flanders and Britain, said that 'we must attract to us everyone in the world who has Nordic blood in his veins, so that men of Germanic or Nordic blood will never fight against us again'. The Nazi principle of 'protecting life by giving death' prevailed as usual. While Poland was the trial-ground of *Todesborns*, institutions for wholesale killing, Northern Europe, Norway in particular, experienced the opposite, the operations of the Lebensborns, the 'sources of life'. No sooner had the Nordic countries been conquered by the sword than orders were given to all German troops to consolidate the victory by the cradle. The results of the campaign for the production of German babies throughout Western Europe exceeded the expectations of even the most optimistic race experts.

The methods used by the Germans in occupied territories are all too familiar. Let us merely recall that after the conclusion of military operations the German administration laid its hands on everything that might be useful or necessary to it. By decrees, requisition orders and acts of submission by the authorities of the defeated country it rapidly extended its control over every field of public and private life, and acted as if the occupation was to last for ever. It is also a familiar fact that the Geheime Feldpolizei, the secret field police, arrived on the heels of the troops to deal with Jews, resistance fighters and anti-Germans. The activities of the Gestapo, the SD, and those who aided and abetted them in every occupied country, generally began modestly and anonymously, only to end in a blood-bath that the people of those countries would not soon forget.

On the other hand, the activities of the RuSHA are relatively unknown to the general public. Whether with the aid of the Wehrmacht High Command or the RuSHA, the Lebensborn organization saw to the establishment in all the occupied Western countries of death's-head maternity homes for the purpose of recovering the precious 'German blood' so urgently needed. The RuSHA applied the principle of intercommunicating vessels, building up in the West the reservoirs of blood that would be required in the East as soon as the latter had been emptied of sub-human blood.

Imagine the situation created by the conquests. In Norway

and Denmark, for instance, fair, blue-eyed, racially unobjectionable children were so numerous, the race so pure, that selection was superfluous. Because of the tremendous value the Nazi race theorists attributed to Nordic women for reproductive purposes, an incredible procreation campaign was immediately directed at the troops who in the spring of 1940 occupied first Denmark and then Norway. The results soon became evident, as is shown by this personal letter addressed to Himmler on 5 December 1940 by the Chief of Police and of the SS of the German High Commission for Norway, SS Gruppenführer Rediess:

In Norway, as in the other occupied territories, a not inconsiderable increase in births must be expected at the end of this year and even more so in the coming spring, the fathers of the children in question being members of the German Wehrmacht, the SS and the police. Though no statistical evidence is yet available in view of the short time that has elapsed since the occupation of Norway, there are increasing signs that the friendly relations established by the German occupation troops . . . have made a substantial contribution to the number of births to be expected. Individual cases are already arising . . . of Norwegian women made pregnant by Germans seeking the aid of the German Reich, above all on the ground that they are despised and boycotted by the Norwegian population because their pregnancy was caused by a German . . .

. . . from the political point of view it should be noted that only a small proportion of the unmarried fathers conclude from the fact of their paternity that they should marry the child's mother and thus bring both mother and child into the sphere of influence of the German Reich. Thus, if we take no steps to care for the far greater proportion of these Norwegian mothers, we shall necessarily be instrumental in increasing the numbers of those who do everything in their power to oppose the German penetration of the Norwegian people. Also these children will be lost to any influence by us, as their illegitimate birth by a Norwegian mother gives them Norwegian citizenship . . .

Politically we obviously expose ourselves to the charge of using the Norwegian people, as it were, as an institution for increasing the number of German births.

. . . to add to the stock of racially valuable blood in our racial community I suggest the establishment of (1) German-controlled maternity homes (Lebensborns) and (2) homes in which mothers and children can be looked after.

This comprehensive survey of the question by SS Gruppenführer Rediess led to prompt action. Maternity homes for the children of German fathers opened one after another throughout Norway. There were nine of them altogether, and they worked full out until the German occupation troops were withdrawn. Estimates of the number of illegitimate births vary between 6000 and 12000; most of the legitimate children of these mixed marriages were born in Germany and not in Lebensborns.

The Norwegian homes were, strictly speaking, not Lebensborn homes, the letter-heading being: 'SS and Police Command at the Headquarters of the Reich Commissioner for the Norwegian Occupied Territories, Lebensborn Department'. However, the Munich headquarters sent SS doctors, midwives, nurses and other RuSHA employees to Norway. In view of the importance of the non-SS contribution (Wehrmacht, Luftwaffe, and so forth) and the purity of the mother's blood, this precious fluid had to be won for Germany at all costs. The Lebensborn organization sent instructions to the Reich Commissioner for Norway on how these homes were to be run, and two prominent race specialists, first SS Sturmbannführer Ragaller and later SS Sturmbannführer Tietgen, were put in charge of them.

A huge dossier survives about the German maternity homes in Norway, as do similar dossiers for Belgium and Holland. There are full records of the names and addresses of the women who gave birth there, their family situation and details of the father, and in fact the Norwegian dossier is the most complete of all those we have been able to consult, whether public or private.

Encounters between local girls and soldiers of an invading army have taken their natural course since history began, and there was nothing exceptionally new or dishonourable about the births of little Teutons in Norway. But the deliberate encouragement of such births for a specific purpose, and the creation of conditions favourable for bringing them about, were Nazi innovations. The intensive propaganda aimed at encouraging the men to have children by these Nordic women and the establishment of institutions designed to relieve them of all material and moral worries in the matter were supplemented by special legislation which, for all its immorality, was essential to the German population policy. In some of the occupied countries the Germans modified the law to abolish parental authority over girl minors, who thus did not need parental consent to marry a German, and in the case of Norway this was the law of 1 February 1941. The Norwegian courts were also deprived of jurisdiction in cases in which a German was involved, which meant that no child of a German father could be adopted by Norwegians without special permission of the Lebensborn.

Many hundreds of Norwegian girls took refuge or were confined in homes in Germany, and most of them abandoned their babies immediately afterwards. Their trail was discovered after the war, as was that of thousands of children kidnapped in eastern Europe, thanks to insurance policies taken out for them by the 'L office' with a Berlin insurance company whose records have survived. Thus we find that in June 1943 152 women and 128 children arrived. In July the number was 156 women and 134 children. After that the numbers increased steadily. From May 1944, when 204 Norwegian women and 199 children arrived at Lebensborn homes in Germany, until the end of the occupation the average was 200 women and 190 children a month.

Because of their high racial quality, Norwegian babies were eagerly snatched up by Nazi families. Unfortunately, as a result of the dirt, disorder and illness that usually prevailed in the homes, there were a number of deaths and, as usual when racially valuable stock was concerned, these cases fill nearly half the files.

The homes in Norway were the subject of a long article in the Swedish newspaper *Dagens Nyheter* on 12 December 1943 under the headline: 'The Lebensborn Society – a new Nazi organization in Norway for the promotion of illegitimate births'. It began by quoting a booklet published by SS Gruppenführer Rediess, only twenty copies of which were published, since it was intended for internal use only: its title was *The SS for Greater Germany – with Sword and Cradle.*

The German Reich [the booklet said] has the right to recognize or reject the German citizenship of all illegitimate children by German soldiers . . .

This people is a Germanic people, and hence it is our duty to educate its children and young people and to make the Norwegians a Nordic people again as we understand the term.

It is definitely desirable that German soldiers should have as many children as possible by Norwegian women, legitimately or illegitimately . . . the combatant soldier must be relieved of all worries that might affect his morale . . . the Reich Commission will meet all the expenses caused by mother and child. The Reich Commission will pay the expenses with bank notes confiscated from or printed by the Bank of Norway.

Dagens Nyheter also gave precise details about the homes. 'A number of luxurious homes for these mothers and children were secured by the requisitioning of Norwegian villas, a number of hotels at Geilo, Hurdal and Aas in the Bergen area, and the Stalheim Hotel, a well-known watering place on the west coast.'

Rediess naturally dealt with the all-important question of the race of the girls who were to have the signal honour of being married to a member of the master race. 'The deciding factor for the approval of a marriage between a German soldier and a Norwegian girl,' he said, 'is not only her purely Nordic external appearance, but the political attitude and way of life of her family.' But the most interesting part of this document concerns the moral aspect:

The problem [of illegitimate children] is often solved by the soldier adopting the child and accepting it in his own family, as German wives show great understanding of the situation and that is also the best solution for the child.

Thus some German wives were so patriotic that they understandingly accepted the little souvenir from Norway that their husband brought with him when he came home on leave. Was this the best solution as far as the child was concerned? The sentence we quoted above merely confirms that children were taken to Germany, with or without their mothers' consent. In fact they were kidnapped, by virtue of the law that prevented Norwegians from intervening in any matter in which a German subject was involved. Thus the child of a German soldier, recognized as valuable by the race experts, was controlled and supervised from birth. not only medically but also administratively. The mother or other responsible person was obliged to apply 'voluntarily' for a certificate of nationality showing that the child was of German stock. In the event of a refusal by the mother or her representative to comply with this formality, a German guardian was appointed who promptly did what was required. In other words, if the mother refused either to go to Germany or to present her child to the German Reich, the child was taken from her on the strength of the certificate that made it a German subject. Sometimes the mother suffered the same fate. For, apart from those who voluntarily handed over their children for adoption or went to Germany either to have the baby or live 'as German citizens with or without a German husband', many Norwegian women were trapped into going to Germany against their will. The kidnapping process was given a semblance of legality by a Nazi ruling that defied the fundamental laws of a sovereign nation and legalized the separation of mother and child against the mother's will. This made it possible to blackmail several hundred Norwegian women who were of 'good Nordic breeding stock' into going to Germany with their babies.

The kidnapping of Norwegian women had been long prepared. In 1940 Ebner wrote to Himmler that it would be highly desirable to bring racially and politically valuable Norwegian women to Germany with a view to the children to be expected. Himmler replied:

> The compulsory transfer to Germany of Norwegian women expecting children by German occupation troops is a unique opportunity. This operation would be in accordance with the objective of transplanting purely Nordic women to Reich territory

in large numbers. There is a special need of 'nordicization' in south Germany.

Rediess's comment was that it was not without importance for future generations that twenty or thirty extra divisions should be made available for the defence of the territories 'conquered by our comrades'.

As always and everywhere, the shadow of the 'L office' lay behind this new tragedy. Ebner was at last promoted to the rank of SS Oberführer, which gave a modest country doctor general's rank; SS Standartenführer Sollmann, only a colonel but a member of the Blood Order, took up his appointment and put his qualifications as a commercial traveller to good use in organizing the recovery of Nordic blood; and Inge Viermetz, former head of the BDM, with her pinched lips and cold eyes, was always in the breach. All three well knew how Rediess and his aides lured Norwegian girls into the trap, with the aid of free cinema performances, concerts in the parks and public gardens, access to the best restaurants reserved for the occupying power, anything that helped innocent girls to fall into the arms of the German fertility gods. All that mattered was the children and their integration into the Reich.

At the other end of the tunnel, Ebner, Sollmann and Viermetz waited patiently for nature to do its work on the Führer's behalf. After that it was only a question of months. At last the telegrams began to arrive:

[Document SS 3046] – Tietgen to Sollmann/Transport. Children arrive Monday. Elbe district has been informed. See to making of further arrangements.

[Document SS 3043] – Nine Norwegian children arrived. Names to follow.

[Document SS 3042] – Dr Peith [woman head of the Sonnenweiss home] to Ebner. Three children in good health, the rest whooping cough. One child Ingrid Wahwik doubtful. Mother should be informed, but as child not adopted [in this context meaning 'alone, unaccompanied by mother'] mother's address unknown.

Faced with an influx of children, many of whom arrived without papers or means of identification, a head nurse, Erika Wittmann, suggested to Ebner in a letter dated 2 September 1942 that 'every arriving child should be accompanied by a personal report'. This suggestion followed the death of a child named Jürgen Hoerst. In its mother's 'absence' it had been impossible to complete the death certificate, and so, to enable the child to be buried, they had had to forge one. The reason for the mother's 'absence' was, of course, that the child was one of the many that were kidnapped.

Here is a 1947 Berlin newspaper cutting:

Norwegian children sought. Seventeen Norwegian families are trying to trace their children taken to Germany by the SS Lebensborn organization and accommodated first in homes and then with families. The Central Youth Office, 1–2 Wilhelmplatz, Berlin, W.8., appeals to the population to help in finding these children. Foster-parents in particular are asked to help.

The children's names follow. After the war the international search services took up the cases of thousands of foreign children kidnapped by the numerous Nazi so-called charitable organizations. The Norwegian children presented them with one of their most difficult problems. As most of the Norwegian children had not been kidnapped but voluntarily offered to the RuSHA by irresponsible young women, the first reaction of the Norwegian authorities was to oppose their return from Germany. In the early post-war years the Norwegian Government actually suggested that illegitimate children might remain in Germany in their own interest. In fact no Norwegian complaint about the actions of the 'race experts' occurs in any of the post-war documents we have been able to consult. This is the more inexplicable in that the Norwegian Minister of Social Affairs had in his possession most of the files about the girl mothers and their children, which were handed to him personally by Ragaller (evidence at Nuremberg trial, January 1948).

The Swedes, however, were more sensitive to the implications of the problem, and immediately organized a rescue operation for the Norwegian children accommodated in reception centres in occupied Germany; many of these children were looked after and subsequently adopted by Swedish families. After long and painful negotiations between the International Refugee Organization and the Norwegian Red Cross, most of the cases were eventually settled, but not always in the child's interests. Though Norway refused to accept 'children of German fathers who might be the bearers of the seed of a new fifth column', as a 1947 report of the International Refugee Organization put it, in the end a number were accepted, and are now, no doubt, fully fledged Norwegian citizens. At all events that is what we assume, for the Norwegian authorities, fearing an irresponsible intrusion on our part into a difficult past, refused to help in our research.

Quisling's Norway occupies the most important place in this chapter on the activities of the Lebensborn organization in Western Europe for a simple reason: nowhere else did the machinery for manufacturing Nordic children for the Third Reich work so successfully. No other country helped the Germans as Norway did, by an agreement, signed by Rismes, the Norwegian Minister of Justice, to legalize the kidnapping of children and their transfer to Germany. So many children were born of German fathers that some had to be placed in Norwegian institutions, and according to the Lebensborn authorities their transfer to Germany became more than necessary. In the course of his interrogation Ragaller claimed that children were sent to Germany only with the consent of the Norwegian Minister of Justice.

This deplorable attitude by a collaborationist government emerges even more distinctly from the documents we examined when we compare it with that of neighbouring Denmark. The attitude of the German invaders of that other Nordic, 'racially valuable' but politically hostile country effectively prevented the establishment of Lebensborn homes there. Children were certainly born of German fathers, and Danes, like Norwegians, Frenchmen, Belgians and Dutchmen, certainly enlisted in the Waffen SS, but the number of children and of SS men 'made in Denmark' never reached the Norwegian proportion.

Nine homes more or less attached to the Lebensborn organization functioned in Norway for nearly five years, while in Denmark things never got beyond the planning stage. Several homes were meant to be established there, but the correspondence clearly shows the difficulties the Germans encountered. In 1944 the NSV, that organization of 'sisters' completely dressed in brown, was still proposing to establish a maternity home as well as a crèche for 'orphaned' children of German fathers ('orphaned' being a euphemism for children of German fathers taken from their mothers to be adopted by German families), but nothing came of it. A section of one clinic requisitioned by the Wehrmacht was reserved for the confinement of Danish women, and Ebner allotted it 120000 marks pending the establishment of a proper Lebensborn. Its trace survives in letters that speak of 'thefts from this clinic' and of the 'permanent concern of the organization to get on well with the local population and the future mothers'. The correspondence ends with the end of hostilities. After the war there is practically nothing about Danish children or Danish women in the files of the International Refugee Organization.

The only British contribution to the Third Reich programme of population expansion was the production of illegitimate children in the Channel Islands. Here the fathers were members of the Wehrmacht, and it was not until May 1944 that an RuSHA commission was ordered to subject to racial examination the illegitimate children and their mothers born in Jersey and Guernsey.

The order stated that 'all the mothers with children already born, in so far as they comply with our racial specifications, will be transferred to Germany' and that all the 'pregnant women will be taken to German maternity homes in France'. To help the Wehrmacht in the matter, according to evidence presented at Nuremberg, the RuSHA also decided to send a 'commando' [*sic*] to the islands, as well as to send 'everything racially valuable' to the Lebensborn organization in Germany.

However, though they formed part of the British people whom Himmler so much wanted his Germans to resemble, not all the products of Jersey or Guernsey were equally valuable in the eyes of the SS. Racially the women there came halfway between the high Norwegian rating and the lower rating attributed to Frenchwomen. The documents point to many instances of mothers being not up to standard. When they were, in most cases the child was taken away and placed in a Lebensborn home somewhere in Europe. The documents show, among other things, how broad was the gamut of methods employed by the RuSHA in laying its hands on children; these varied according to requirements, but the mother's feelings were never taken into account.

On 8 June 1944 – just two days after the Allied landing in Normandy – Yvonne S., a young woman of French origin but British nationality, made an application for German nationality. Because of her antecedents she was classified in category 3, that of women who were 'undesirable', meaning racially undesirable. The examiner made the following report:

Yvonne S. is expecting a second illegitimate child by the same German, Horst B. Nevertheless I regret not being able to place her in category 2 (possible), the lowest necessary for the granting of German nationality. In spite of everything the girl makes a good impression. I therefore ask the leader of the RuSHA for a decision in relation to her and her children, who could be transferred to a Lebensborn home.

What happened to Yvonne? Did the father of her two children marry her before the 'L office' commando grabbed them? At St Helier there are a number of people with a name strangely resembling hers, but none of them have ever heard of her. The two illegitimate children of a little French girl who vanished without leaving an address were not the only living witnesses to a German occupation which did not have only enemies, after all. Hence this secret letter, dated 24 May 1944, sent by the head of the race office in France (signature illegible) to headquarters in Berlin:

According to a communication from Area Command no. 515

at St Helier, since the occupation of the Channel Islands between sixty and eighty children have been born whose fathers are unquestionably members of the German occupying forces. In view of the high Nordic component in the race of the island population, which is substantially higher than that of the population of continental France, the mothers and children concerned should mostly be racially valuable and flawless, so the RuSHA leader considers that the racially and genetically flawless children, and perhaps also mothers, should contribute to the desired growth of population of the German people by being resettled in the Reich.

The situation of these unmarried mothers is very bad indeed, as on the one hand they are persecuted by the rest of the civil population and on the other have been refused any maintenance allowance for the children by the military command in France (justice department). In most cases these English [*sic*] mothers have learnt German, so that they could be transferred to Germany without difficulty.

At a conference on 24 May 1944 between the Lebensborn representative and senior officer of the SS and police in France, SS Sturmbannführer Dr Fritze, and me, it was decided that in future the Lebensborn should take over responsibility for the expectant mothers in these islands after previous investigation by the RuSHA leader.

It would be a mistake to conclude from these documents that Himmler's ambitions in relation to the transfer of population were limited to children fathered by members of the Wehrmacht or the SS. His intentions were quite specific. Belgium, Luxembourg, Holland, Alsace-Lorraine, Norway and Denmark were to be incorporated into the Reich. The British were to be transplanted to the Baltic. Burgundy was to be detached from France to form the basis of a future SS model state, the Regent of which was to be Heinrich Himmler and the Chancellor the Belgian SS leader Léon Degrelle.

In March 1943 Himmler said:

After the conclusion of peace the world will discover that the Burgundy of old, that country which was above all other things a land of science and the arts and was reduced by France to the level of an appendix preserved in cheap wine, has been resuscitated. The sovereign state of Burgundy, with its own army, its own laws, its own currency and postal service, will be the SS model state. It will include French Switzerland, Picardy, Champagne, the Franche-Comté, Hainaut and Luxembourg. The official language will of course be German. The National Socialist Party will have no influence. Only the SS will rule, and the world will look with astonishment and admiration at a state in which the ideology of the SS has been put into practice.

12. Bastards of Good Race

'Soldiers on leave must be offered good company, and this for reasons of population policy.'
MARTIN BORMANN, Circular 83/44

The amount of fraternization that took place in various countries can be judged by comparing the approximate figures of births 'by German fathers' that took place in each, proportionate, of course, to their population. After the 'plus' example of Norway and the 'minus' one of Denmark, we turn our attention to Holland, where the Lebensborn organization ran the Gelderland home, near Nijmegen, for several years. Unlike the homes in Norway, which accepted the children of German fathers in general, this home was reserved for young women made pregnant by members of the SS. How many Dutch women passed through this home, as unmarried mothers or as the legal wives of SS men, is unknown, but it can be safely assumed that there were several hundred. In Holland, as in Denmark, in spite of the efforts of pro-German circles to evade or modify the law on the protection of female minors and their illegitimate children, the hostility of the population to girls who slept with Germans was such that the 'dealers in Aryan blood' were forced to exercise restraint. (Children of former collaborators are still boycotted in some Dutch villages.) Hence the innumerable letters, circulars and orders in the 'L office' files recommending sustained efforts to establish good relations between the occupying authorities and the local population, at the same time emphasizing the necessity of leaving real Germans in positions of responsibility at the Gelderland home (as medical superintendent, head nurse and so on).

Some Dutch women evaded obloquy by going to France, Belgium or Germany to have their babies. One such was Baroness Juul op ten Noord, who was headmistress of the Reich School for young ladies at Hejthuijsen and a personal friend of Himmler's, and from the racial point of view constituted a special case. She received the following letter dated 15 September 1943:

Dear Juul, I was delighted to see you again in Berlin after such a long time. To avoid confusion and to prevent anything going wrong, I am confirming in writing the arrangements we then made . . . You will go to Berlin on 1 October. SS Obersturmbannführer Baumert, whom I have taken into our confidence, will see to all further arrangements for you. You must tell us yourself what German name you are going to choose. Obersturmbannführer Baumert will have a pass prepared for you in which you will be described as a German married woman born in the Netherlands and married to an agriculturist . . . I suggest that you go to one of our Lebensborn homes a month or two before your confinement and await your baby there. Both the registration of the birth and the fact that you have had a baby will be kept secret . . . The registration will be seen to by my staff. . . Heil Hitler!

(Signed) H. Himmler*

All the usual 'L office' administrative features are present in this instance – false name, false pass, secret registration of the birth at the Lebensborn's own register office. Also Himmler advises the baroness to abandon and subsequently to adopt the child, and to tell it the whole story eventually, 'when the right time came'. The father was not a German, but a Dutch member of the SS who seemed so dilatory about doing the right thing by her that on 27 October 1943 Brandt, acting on Himmler's instructions, wrote to him very sharply indeed, saying, among other things:

The Reichsführer SS calls on you to revert to the standard of behaviour to be expected of a Germanic male in affairs of the heart and relations with women.†

We quote this case of a child born of a 'foreign father' in a Lebensborn because it was apparently unique in the annals of the 'L office'.

Immediately after the war it seemed legitimate to assume that the Norwegian, Dutch, French and Belgian children

*H. Heiber, *Lettres de et à Himmler*, Paris, 1969.
†H. Heiber, *op. cit.*

found in Germany (about 50000 from Holland, 40000 from Belgium, 20000 from France and 12000 from Norway and Denmark) were fathered by SS volunteers who were natives of those countries. But investigation quickly showed that the recreation of the warrior organized by the Lebensborns was the only privilege reserved to the very end for German SS men, or, at any rate, super-Germans of the Wehrmacht. It was thus solely in her quality as a personal friend of the Reichsführer's that Baroness Juul was granted the signal honour of giving birth in a death's-head maternity home to a child neither of whose parents was German. She was confined at Steinhöring, and left behind a good impression among the nurses.

The documents dealing with the Ardennes home, at Végimont, near Liège, in Belgium, contain nothing but complaints and criticisms about Belgium in general and of the Belgians employed there in particular. According to the correspondence between the home and the headquarters in Munich, the Belgian staff – midwives and nurses – did everything possible to sabotage the production of little Aryans. Ebner, in a letter describing the ill-will of the Belgian staff, wrote to a friend that he was 'depressed' at being unable to find the good doctor so urgently needed for the Ardennes and Gelderland homes.

Early in 1943, a seven-month-old baby of German parentage on both sides died suddenly in the Ardennes home. This put all the Germans in Belgium in a turmoil, for rumour had it that neglect by a Belgian midwife was responsible. The Gestapo carried out a rapid inquiry, which ended with the dismissal of the midwife, Fanny M. A few days later a post-mortem showed that the cause of death was a brain tumour. The incident well illustrates the climate of mistrust that existed. SS Sturmbannführer Lang, the head of the home, wrote to Ebner:

I have informed Comrade Friedrich that we should not drop Sister Fanny because, as a result of her work for Germany, she is maligned by her Belgian fellow-countrymen and will never again get a job in any Belgian institution* ... There is a great deal that could be said about the Belgian homes. We are not convinced that the children accommodated in them get everything they need. It has been reported to me that a hostile political attitude plays a big part in this.

The Ardennes home, which was intended to meet the needs of Belgium and Northern France, was opened at the beginning of 1943 by Inge Viermetz, who for some time acted as head of the home herself. Belgian girls, most of them Flemish, were confined there, as well as some from Holland and France. The SS took careful precautions to ensure the secrecy of the births there, as is shown by the expectant mothers' continual comings and goings between homes far from where they lived. Ebner visited the place in November 1943, and found nothing but 'dirt, slovenliness, and unqualified personnel, not even able to speak German'. According to Sollmann, who visited the home a few weeks later, matters were still worse. On 30 November 1943, Ebner wrote to SS Sturmbannführer Lang:

SS Standartenführer Sollmann has told me that the midwife at your home uses a so-called delivery chair on which she places the mother shortly before the birth. This chair is said to be in the delivery room. During my last visit I did not notice it, and I do not recall any mention of that method of delivery. I must ask you to have this chair removed from the delivery room immediately. It goes without saying that women in the Ardennes home will be delivered on a delivery bed, just as they are in other Lebensborn homes. Sister Hanna will be arriving to take over as midwife in the next few days, so that in future everything will be in order in this respect.

On top of all the other problems, the hostile attitude of the Belgian population, the animosity of the nuns of the Bavaria hospital, near Liège University, towards the Lebensborn children, the shortage of midwives, nurses and gynaecologists, and a diphtheria epidemic which cut the home off from the outside world for weeks, a plague of mosquitoes sud-

*There is evidence that this Sister Fanny later worked in the French Lebensborn, where she remained until the German troops withdrew.

denly arose. To cope with this new menace, SS Sturmbannführer Lang wrote to Steinhöring in July 1944 for a pamphlet entitled 'The Biggest Mosquitoes in the World', and Ebner conscientiously forwarded the letter – to Auschwitz, where scientific experiments in combating insects were carried out in the concentration camp.

But before a reply could come from Auschwitz enemies more difficult to destroy than any insects arrived at the gates of Liège. The home was hurriedly evacuated in a few hours, and the Allied soldiers who arrived found it empty and the documents burnt or vanished. The mothers had also vanished into thin air, leaving their babies to Lang, who took them by road to Schalkhausen, near Ansbach, in Bavaria. Four of them were only a few days old. The children, after a journey of several weeks, finally found refuge at Steinhöring. The precious blood had to be rescued at all costs, and the obvious place to take it was Bavaria, the cradle of Nazism and the stronghold of the Lebensborns, where Ebner was waiting.

The basic principle of the SS, and of the RuSHA in particular, was racial purity, the superior biological value of the men, women and children called on to participate in building the Germany of the future. Thus the SS leaders decided at the outset to restrict the privilege of wearing SS uniform to Norwegian, Swedish, Danish, Dutch and Flemish volunteers. This continued so long as the success of German arms permitted them to be finicky, and French, Walloons, Spaniards and Croats who wanted to fight under the swastika had to be content with enlistment in the Wehrmacht. Not being of German blood, they could only taint the Nordic race after final victory.

With the decline of German military fortunes, bigger and bigger inroads were made into the principles of selection that governed admission to the SS, and soon the characteristics of an élite that Himmler wanted to give it were no more than a memory. By 1944 the strength of the Waffen SS was about 800000 men, of whom more than half were foreigners, most of them 'not racially valuable'. Ideological principles, flouted by the presence of Muslims, Cossacks, Ukrainians and other 'sub-humans' whom the Reichsführer SS so despised, quickly receded into the background, and the blood spilt in the defence of the tottering Greater German Reich changed colour. The colour of the hair and eyes of the combatants grew darker until it was sometimes an 'oriental' brown, and it became rarer and rarer to see a crop of fair hair standing out in the columns of harassed, retreating troops. After Stalingrad, 'Nordic man' no longer had to be at least six foot tall, or 'walk like a king', or even have 'a bright, victorious look'.

The requirements of the race experts in the matter of child selection also varied with the ups and downs of an increasingly difficult military situation. The severity of the fighting on the Russian Front resulted in losses on a scale that the *Blitzkrieg* of 1930–40, both in East and West, had not led the German leaders to foresee; and though within the Reich planned propagation by SS men and selected girls continued, the recovery of Germanic blood in the occupied countries in the East ended by becoming much more a question of quantity than of quality.

The most revealing evidence of this new turn in the theory of the super-race, which would have been inconceivable a few years earlier, is a letter to Himmler dated 29 May 1942 from Dr Conti. This tireless campaigner for artificial insemination, which he regarded as the only method by which pure-blooded future generations could be produced in sufficient numbers, was now actually willing to contemplate the use of Frenchwomen to help repopulate the Greater Reich:

> Reichsführer,
>
> In France altogether about 50000 children have been born of Frenchwomen. In my opinion these children are not bad, most of them not worse than those produced in Norway by Norwegian women . . . I suggest that the Lebensborn should intervene energetically to secure these children.

A few months later, in October 1943, a high official in the German High Command (signature illegible) stated that, according to a confidential communication from the (German) medical superintendent of the hospital at Suresnes. near Paris, 'the number of illegitimate children born in France has now reached 85000'. The growing interest in the production of children by French mothers (by the end of the Occupation the number seems to have exceeded 100000) concerns us here only to the extent of its relevance to Nazi racial policy.

Up to May 1942 the children of the Occupation in France were in a position similar to that of French volunteers for the German army; it was not considered necessary to gather statistical information about the former, and the latter were not thought worthy to wear SS uniform. The French were considered to have been 'bastardized' by mixture with Arabs, blacks, Jews, and so on. In Himmler's opinion this 'bastardization' was responsible for France's military and political weakness. He nevertheless gave orders towards the end of the war that steps should be taken to track down French officers with German-sounding names with a view to taking over their offspring. After the war the authorities found out that 35000 children had disappeared from France without a trace.

The tentative interest that the Nazis took in French children born of German fathers and the enlistment of foreign volunteers, including Frenchmen, in the Waffen SS were the premonitory signs of the relaxation of the rules of racial selection that circumstances imposed. Recruits whose ethnic origin would have made the race experts' hair stand on end a few years previously in 1943 formed an integral part of what had been the purely Nordic SS. The Frenchmen of the Charlemagne Division, like the Muslims of the Handschar Division or the Ukrainians of the Galicia Division, were in fact mere sub-humans promoted to the temporary rank of first-class human beings for the sole purpose of keeping the German war machine in motion.

In all the documents we examined we found no information whatever about what happened to the illegitimate French children of German fathers. It is nevertheless known that most of them remained with their mothers, Conti's proposals not having been followed by any large-scale operation to transport them to German territory, as happened with the Norwegian children. After the war the French Red Cross, like the Red Cross of all the European countries, took action on their behalf. The Z service, specially established for the purpose, granted paternal rights to the Red Cross, above all to avoid dramas with returning prisoners of war whose families had increased in their absence without their knowledge.

We had the greatest difficulty in tracking down the site of the one Lebensborn maternity home in France. In the documents it appears only as the Westwald home, which was what the SS called it. Westwald means 'Western Forest', so we naturally searched for it in the forests west of Paris, at St-Germain-en-Laye and Versailles. Eventually we realized that, from the SS headquarters in Berlin, it was the city of Paris and not the forest that was in the west.

The Lebensborn home at Lamorlaye, near Chantilly, north of Paris, was opened with the usual SS ceremonial on 6 February 1944, in a splendid requisitioned château that had belonged to the Ménier family and today belongs to the French Red Cross, which has turned it into a magnificent re-education centre for handicapped children. It had at first been occupied by the Wehrmacht (cavalry), but the SS took it over in 1942. It was actually functioning earlier than 1944, several children having been born there the year before. The officer in charge was SS Oberstabsarzt und Sturmbannführer der Schutzpolizei Dr Fritze, who now has a practice in a town in North Germany. The choice of such a man as head of a maternity home may seem surprising, and his conduct was the subject of perpetual criticism in the 'L office', for he preferred the gay life of the Paris restaurants and night spots to bothering about the home at Lamorlaye, to which he paid only occasional visits, generally in a state of 'advanced drunkenness'.

However, on 15 March 1944 he found time to write to Dr Ebner, asking for help 'in our struggle against flies and mosquitoes'. A letter of Ebner's dated 2 May states:

> . . . the Westwald home near Paris is especially short of matches and requests you to make a substantial delivery of them. The home is also short of provisions of all sorts, such as semolina, rice, porridge and cocoa. Also some packets of coloured pencils are urgently needed in the home.

In view of the situation on the Western Front, this appeal remained unanswered.

Who were these girls who in May 1944 still lived on a diet that included cocoa, porridge and semolina? And who were the fathers of the children born in the Westwald home? The answer to the second question is easy: all the fathers of the children born at Lamorlaye belonged either to the SS or the police. Since Chantilly is a famous horse-breeding area, the German cavalry had set up its headquarters there, and remained there from the beginning to the end of the Occupation. No SS detachment was ever stationed in the area; consequently the fathers of these children must have belonged to various SS units stationed all over France, and particularly in the Paris area. The Wehrmacht cavalrymen had no more right than the local inhabitants to penetrate the château precincts, which were guarded night and day by armed sentries. As an additional precaution, the Germans dismissed the French staff, including the gardener, 'who in any case was not fond of the *boches*'.

Needless to say, recourse to official sources of information was denied us, but painstaking 'semi-official' inquiries eventually resulted in our finding the trail of the young women who passed through this home to offer their child to the Greater German Reich. They were French, Belgian, Dutch or Norwegian. Most of them abandoned their babies to the Lebensborn authorities, who evacuated them in August 1944. In 1945 they were found still at Steinhöring as there had been no time to have them adopted by German families. The German documents about them enabled the allies to trace their mothers, or, at any rate, a large number of those who had had their babies in the home under their real names.

On 13 November 1944 Ebner wrote to S S Obergruppenführer and Police-General Rösener: 'We have unfortunately had to evacuate our homes in Holland, Belgium, Luxembourg and France. It is to be hoped that the time will soon come when we shall be able to move back into them.'

PART TWO

The Orphans of Hate

'The world will not dry its tears until the last stolen child has been restored to its real family.'

Revue mensuelle des Questions allemandes, April 1949.

13. 'Fountains of Death'

> 'All the good blood in the world, all the Germanic blood that is not on the German side, may one day be our ruin. Hence every male of the best Germanic blood whom we bring to Germany and turn into a Teutonic-minded man means one more combatant on our side and one less on the other. I really intend to take German blood from wherever it is to be found in the world, to rob and steal it wherever I can.'
>
> HEINRICH HIMMLER, speech to officers of the Deutschland Division, 8 November 1938

The word *Lebensborn*, 'fountain of life', was a Nazi invention, but reading the documents about the Germanization of the occupied territories in Eastern Europe suggests that *Todesborn*, 'fountain of death', would have been a more appropriate neologism. This applies particularly to Poland, the four western provinces of which were declared to be German by decision of the Führer dated 8 October 1939 and placed under the so-called General Government headed by Hans Frank.

The process of Germanization is best illustrated by some chronological details.

On 15 March 1940 Himmler ended a speech to the commandants of concentration camps with the words: 'Then all the Poles will vanish from the face of the earth.'

On 15 August 1940 he addressed a top-secret six-page memorandum to the Führer, entitled 'Some Thoughts on the Treatment of Foreign Populations in the East', and informed his confidants shortly afterwards that the document had been read and approved. Also, to ensure the utmost secrecy, he had ordered that very few copies were to be made. The head of the RuSHA and of the Lebensborn Society were among the privileged recipients.

The document begins with a reference to various racial groups in Eastern Europe other than the Poles and the

Jews, including the Ukrainians, White Russians, 'Gorales', 'Lemkens' and 'Kashubs'. Such small racial groups were to be recognized and encouraged, he said, and if any other such small groupings should be discovered, they would apply. By this he did not mean that any unification of the population in the East should take place; on the contrary, it should be split up and subdivided to prevent the development of any national consciousness and culture. The object was to 'break them up into innumerable little fragments and particles'. Members of these small 'tribes' might be used as police officials and burgomasters.

There must be no grouping at the top, as only by dissolving this whole ethnic brew of fifteen million in the General Government and eight million in the eastern provinces shall we be able to carry out the racial sifting which must be the basis of our considerations and fish the racially valuable elements out of this brew and take them to Germany to assimilate them.

The document continues:

I hope to see the very idea of 'Jew' completely wiped out by the possibility of a great emigration of all Jews to Africa or some colony elsewhere. It must also be possible in a somewhat longer period of time to make the ethnic notion of Ukrainian, Gorale and Lemken disappear from our territory. Everything we have said about these splinter races applies on an even larger scale to the Poles.

A fundamental question in the solution of all these problems is the school question, and with it the question of the screening and sifting of the young. For the non-German population in the East there must be no schooling higher than the fourth elementary grade. The objective of this elementary schooling must be merely [to teach children] simple addition up to 500 at the most, how to write their name, and instruct them that it is a divine commandment to obey the Germans and be honest, industrious and upright. I do not think reading is necessary.

Otherwise, Himmler said, there were to be no schools in the East. Parents who wanted a better education for their children, at the elementary stage and later, would have to submit an application to the senior leaders of the SS and the police. The decision would depend primarily on whether the child was racially 'perfect'. If such a child were acknowledged as being 'of our blood', the parents would be informed that the child was to be sent to school in Germany to remain there permanently.

However cruel and tragic this might be in individual instances, this method, if the Bolshevik method of physical extermination of a people is rejected as un-Germanic and impossible, is yet the mildest and the best.

The parents of these children of good blood would then be faced with the choice of either surrendering their children – after which they would probably have no more, with the resultant disappearance of the 'danger' involved in the development of a 'caste of leaders of blood equal to our own – or they will be forced to go to Germany and become loyal German citizens. . .'

In the course of the next ten years, the population of the General Government will consist of a permanently inferior population, to which there will be added people from the Eastern provinces as well as all the areas of the German Reich who share the same racial and human characteristics (some of the Sorbs and Wends, for instance). This population will be available as a leaderless race of labourers and will provide Germany yearly with migrant workers and workers on special projects (road building, quarrying, construction work); they will have more to eat and will lead a better life than under Polish rule, and with their lack of culture, and under the strict, consistent and just leadership of the German people will be called on to co-operate in the latter's everlasting cultural achievements.

The Reichsführer issued the following circular on 14 June 1941:

1. I think it right that young children of especially good race belonging to Polish families should be gathered together and brought up by us in special, not too big crèches and children's homes. Health reasons should be given for taking the children away.
2. Unsuitable children should be returned to their parents.
3. I suggest that for the time being we start with only two or three such establishments, to enable us to gain experience.
4. After six months the family tree and origins of children who

seem to be turning out well should be obtained. After a year consideration should be given to handing over such children to be brought up in childless families of good race.

5. Only the best and racially most enlightened men and women should be appointed to run such institutions.

A top secret order, no. 67/1, was signed in the winter of 1941 by SS Gruppenführer Ulrich Greifelt, head of the Central Office of the SS and SD in Poland:

Re Germanization of Children from Polish families and from former Polish orphanages.

There are a large number of children in former Polish orphanages and with Polish foster-parents who by reason of their racial appearance should be regarded as children of Nordic parents . . .

So that children whose racial appearance leads to the assumption of Nordic parentage may be brought back to Germanity, orphans in former Polish orphanages and in the care of Polish foster-parents must be subjected to a process of racial and psychological selection. The children who are recognized as bearers of blood valuable to Germany are to be Germanized.

The children who on the basis of the racial and psychological selection procedure are classified as being suitable for Germanization will therefore be accommodated between the ages of six and twelve in state boarding schools and between the ages of two and six with families to be indicated by the Lebensborn Society . . .

The directive included the following:

II. 1. My representative will inform the Lebensborn Society of the children aged between two and six who have been recognized as being capable of Germanization. The Lebensborn Society will in the first place transfer the children to one of its children's homes. From there the Lebensborn Society will see to the distribution of these children among the families of childless SS men with a view to subsequent adoption. The Lebensborn Society will assume guardianship of the children accommodated in Lebensborn children's homes. . . .

3. These children are to be treated as German children even before the granting of German nationality.

4. All children now in former Polish orphanages will be sifted and accommodated first. Children living with Polish foster-parents

will be dealt with after the conclusion of this operation. To avoid any unrest among the Polish foster-parents, it must be explained to them so far as is practicable that the children are to be placed in schools or convalescent homes as the case may be. . . .
III . . . 2. I request the Lebensborn Society and the Inspector of German boarding schools to send me comprehensive half-yearly reports (beginning on 1 September 1942) on the progress and conduct of the children entrusted to them.
3. Particular care must be taken to ensure that the term 'Germanizable Polish children' does not come to public knowledge to the children's detriment. The children should rather be described as German orphans from the regained Eastern territories.

A further document, dated 17 September 1942 and addressed to senior officers of the SS, chiefs of police, senior SS officers concerned with matters of race and population and heads of offices outside the German frontiers, stated:

By agreement with the head office of the Reich Commissioner for the Consolidation of the German Race and the Lebensborn, the RuSHA is the competent agency for the Germanization of the names (forenames and surnames) of Germanizable parentless children. This Germanization is to be carried out by the SS leaders concerned with racial and population matters and/or the heads of offices outside the German borders engaged in the process of racial selection. The Germanization is to be carried out in such a way that the new names are as close as possible in root and sound to the previous names. If a Germanization of the previous name is not possible, a new German name is to be given. German names in general use are to be selected (no names with denominational implications, of course); the use of strikingly Nordic names is to be avoided. The children are to be transferred to homes or Lebensborns or boarding schools as the case may be under their Germanized names.

On 30 June 1942 Himmler sent the following message to SS Gruppenführer Friedrich-Wilhelm Kruger, senior SS officer and police chief at Cracow:

Sexual intercourse between a member of the SS or the police and a Polish woman is to be regarded as an act of military disobedience and dealt with by court-martial . . . I will permit mere disciplinary punishment only in quite exceptional circumstances, e.g. in the case of a very young ethnic German who has grown up in Poland and after only brief membership of the SS or the police has not yet learnt to keep the necessary distance from the Polish population . . .

I am not unaware of the difficulties regarding sex that exist for members of the SS and the police in the General Government. I therefore have no objection to sexual relations in brothels or with professional prostitutes who are under administrative and medical control, as in this instance neither the production of offspring nor the establishment of any emotional tie is to be expected . . .

In the course of his 'table talk', on 22 July 1942,* Hitler said that

If any such idiot tried to put into practice such an order [forbidding abortion] in the occupied Eastern territories, he would personally shoot him up. In view of the large families of the native population, it could only suit us if girls and women there had as many abortions as possible. Active trade in contraceptives ought to be actually encouraged in the Eastern territories, as we could not possibly have the slightest interest in increasing the non-German population.

As a loyal follower of the Führer who had personally charged him with purging the Eastern territories of harmful elements, Himmler repeatedly issued directives to his race experts, stressing the same points about abortion and birth control. Intensive propaganda was to be directed to persuading these people that having many children was harmful; that every birth endangered the mother's life; that abortion was healthy and therefore advisable, and would take place under scientific supervision; and that it was desirable for women to submit to voluntary sterilization.

Addressing a meeting of generals at Bad Schachen on 14 October 1943, Himmler said:

It is obvious that in this mixture of peoples some very good racial types will appear every now and then. I think it our duty in these circumstances to take these children, even if we have to

*Picker, Henry (ed.), *Hitlers Tischgespräche im Führerhauptquartier, 1941–42*, Bonn, 1951, quoted in Ackermann, Josef, *Himmler als Ideologe*, Göttingen, 1970.

rob or steal them. This may painfully affect our European sensibility, and many people will say to me: How can you be so cruel as to want to take away a child from its mother? To that my answer is: How can you be so cruel as to be willing to leave a brilliant future enemy on the other side who will kill your son and grandson? Either we recover this superior blood and use it ourselves or – this may seem cruel to you, gentlemen, but nature is cruel – we must destroy it. We cannot take the responsibility of leaving this blood on the other side, enabling our enemies to have great leaders capable of leading them. It would be a crime if the present generation hesitated to make a decision and left it to its descendants.

On 14 January 1944 Hans Frank, head of the General Government in Poland, said at Cracow: 'Once we have won the war, all these Poles, Ukrainians and anything else that may still be running around here can be turned into mincemeat as far as I am concerned, unless anyone has any other ideas for them.' This same Hans Frank, referring to the habit of some German commanders of posting up on the walls lists of persons condemned to death as enemies of the Reich, said: 'In Prague, for instance, big red posters were put up saying seven Czechs had been shot that day. I said to myself that, if we had to put out a poster for every seven shot Poles, the forests of Poland would not be sufficient to provide the paper for such posters.' This was reported in the *Völkischer Beobachter*, 6 February 1940.

14. The Kidnappings

> 'In the eleven years during which I have been Reichsführer SS my aim has always been the same: to create an Order of good blood that can serve Germany and commit itself without sparing itself; to create an Order that will so spread the idea of Nordic blood that we shall attract to us all the Nordic blood in the world; to deprive our enemies of that blood and annex it to ourselves. We must take it for ourselves, and the others must have none.'
>
> HIMMLER, addressing the officers of the Liebstandarte Adolf Hitler, 7 September 1940

Speaking on 4 October 1943 to the heads of the agencies directly concerned with kidnapping children, the Reichsführer said:

. . . A member of the SS must be decent, fit, loyal and a good comrade to his fellows, but not towards representatives of other countries. For instance, he is not interested in the fate of a Russian or a Czech. We take those among these peoples who are of good blood, we steal them just as we steal their children, and we shall see to their education. The conditions in which these peoples live, or whether things go well or badly with them, are a matter of complete indifference to us. They interest me only to the extent that we need them as slaves for our culture. Otherwise they do not interest me at all. If 10000 Russian women die of exhaustion in digging an anti-tank ditch, that interests me only to the extent that the ditch is ready for Germany. We Germans, who are the only people in the world who have a correct attitude towards animals, will also have a correct attitude towards these animal human beings.

A large number of agencies were involved in the kidnappings. Among them were:

The Volksdeutsche Mittelstelle (office for the repatriation of ethnic Germans), known as VoMi, which was responsible for deportations, forced evacuations, kidnappings properly so called, and forced labour. It had its own transit camps and worked in close co-operation with the Lebensborn.

The Nationalsozialistische Volkswohlfahrt (Nazi Welfare Association), known as the NSV, which generally dealt with children too old for the Lebensborn homes, but with the same objectives.

The Reichssicherheitshauptamt (Reich Security Office), or RSHA, which 'took charge' of potentially dangerous elements, that is, it saw to the extermination of Jews, Poles and Russians.

The RuSHA, which was responsible for racial selection and selected children for kidnapping or extermination. It also selected candidates for concentration camps, sterilization and abortion.

The Reichskommissariat für die Festigung deutschen Volkstums (Reich Commission for the Consolidation of the German Race), or RKFDV.

The records of the Nuremberg war crimes trials also contain evidence of Wehrmacht complicity in the kidnapping of children in Eastern Europe. They refer among other things to the seizure of between 40000 and 50000 children by Army Group Centre and of more than 50000 in Ruthenia (Hungarian Ukraine) who were taken to Germany, as well as to Operations Hay and Gypsy Baron (which involved the abduction of more than 100000 adolescents for forced labour in Germany) and the evacuation of Zamosc, with which we shall deal later. Moreover, a number of those involved – now adults long since repatriated to their own countries – confirm that they were taken to Germany by the Wehrmacht. We mention this Wehrmacht involvement in the recovery of children who were 'valuable' for the purposes either of Germanization or enslavement only to give a better idea of the scale of the measures taken to lay hands on children of other nationalities with a view to progressively reducing their biological potential. Only defeat prevented Germany from pursuing this aim.

Because of its specialization in the matter of Nordic breeding, the Lebensborn organization was the obvious agency for Germanizing the children abducted from Eastern Europe. The programme was initiated as early as 1940, and the new task led to several weeks of feverish activity at the Munich headquarters. Sollmann held meetings much more frequently than usual with his principal advisers – Tesch, Ebner, Inge Viermetz and Bartel, his representative at Poznan – and it was decided, in agreement with the Reichsführer, that it was preferable for the organization to deal with children under six. There was a simple reason for this: whether Polish, Russian or Yugoslav, at this age they would be more receptive to Nazi indoctrination than the older children entrusted to the NSV or the VoMi. Because they were so young, they would remember so much less. which would enable Dr Tesch, the Lebensborn legal expert, to falsify their identity the more completely; the susceptibilities of future adoptive parents, who might perhaps turn up their noses if allotted a child of inferior breed, had also to be borne in mind. The honourable Nazi citizens concerned would be told that the child's real parents had been killed in an air raid at home or in the course of military operations abroad. By 1941 in Germany, Party and SS members were falling over themselves in their wish to adopt a child of good blood and so give pleasure to the Reichsführer, and so demand had outstripped supply. Within a few months the round-ups of children in the occupied territories would make it possible to satisfy the demands of childless couples.

Sollmann considered it right and natural that his organization should be chosen to deal with these children, but the Lebensborn homes, established for the reception of girl mothers and their illegitimate babies, obviously could not accommodate all the children who started arriving in ever larger numbers from Kalisz, where there was a big transit camp at which sorting took place. The homes at Bad Polzin and Wiesbaden were turning clients away, so new homes had to be opened to cope with the influx. Kidnapped children were given German names as soon as they arrived at the transit camp, where they were taught the German language and Nazi ideology. The boys spent a part of the morning learning to march in step, salute the flag, give the Hitler salute, and recognize SS and Wehrmacht uniforms and ranks. Everything possible was done to Germanize them as thoroughly as possible. One of the primary conditions for the success of this operation, which generally lasted six months, until they were adopted by German families, as the breaking of all contact with their Polish parents as well as with the local Polish population. A decree was published in German and Polish forbidding the adoption by Poles of children selected by the Germans, and Polish religious establishments were forbidden to 'keep biologically healthy children'. These institutions, which had so far continued to function normally under the German occupation, were often closed, then requisitioned. Priests, nuns and non-valuable children were expelled and sent by force to areas reserved for slaves – if they escaped extermination in the concentration camps.

The methods of Germanization, like those dealing with the children of planned reproduction, were based on Nazi racial theories. It was not now a matter of selecting a man and woman before encouraging them to mate, but of selecting an already existing product, a living child. Doctors specializing in 'racial knowledge', all members of the SS or the police, were put in charge of racial testing at the reception centres. The largest, capable of receiving or selecting several hundred children a day, were at Poznan, Pushkau, Brockau, Kalisz, the Pommern home at Bad Polzin (which was also a Lebensborn maternity home) and in the Gostynin area. But sorting centres were hastily established all over Poland and later in Russia, often with the aid of the Wehrmacht. The children's heads, bodies, arms and legs were measured, as well as the pelvis in the case of girls and the penis in the case of boys, and they were then divided into three groups: (a) those representing a desirable addition to the German population; (b) those representing an acceptable addition to that population; and (c) the unwanted.

To separate the wheat from the chaff as quickly as possible, selections took place annually in all the occupied territories.

More than 200000 Polish children* were thus declared by the RuSHA to be 'racially useful'. The 'useless', including the Jews, who had no right to any selection, exceeded two million, of whom 1800000 were under the age of six.

Himmler had said that the colonization programme in Poland must be carried out pitilessly, so that these Eastern provinces might again become German and be settled by fair-haired Germans. But before the children concerned were sifted by Heinze, Bartel and Viermetz, before they were allotted a phoney birth certificate concocted by Dr Tesch, before they were examined by Dr Ebner, distributor-in-chief of new dates of birth, before they were handed over to good but childless Nazi couples by SS Standartenführer Max Sollmann in full uniform, these children had had families and homes of their own.

Who first spotted and then abducted them? As is usual in the case of the Nazi apparatus, it is hard to give a definite answer. Every member of the SS, the police or the Gestapo, every RuSHA medical adviser, and everyone who worked for any other major Nazi agency was expected to work for the cause. In a country where fair hair and blue eyes are far more common than in Germany there was plenty of choice. It was only necessary to eliminate 'prominent cheekbones, protruding ears or excessively thick lips'. Those who conducted the hunt for future little Germans looked to twelve different sources for their prey: all places where children were assembled; children of Polish adoptive parents or unmarried mothers; children having Polish guardians; children of mixed (Polish-German) marriages; children whose parents opposed Germanization; children of mixed marriages whose parents had divorced; children of deported, liquidated or banished parents (the great majority); children

*It should be borne in mind that before the war the percentage of children in the Polish population was higher than in any other European country except Russia The percentages between 1930 and 1936 were Poland 43; Bulgaria 38; Rumania and Italy 37; Britain and Switzerland 33; France and Germany 30. After the war the figure for Poland had fallen to 35 per cent for a population reduced by six million. These figures, published in London in 1946, show that nearly two million Polish children had lost either father or mother or both and that nearly five hundred thousand children had wholly or partially lost trace of their family.

picked up at random; children born in concentration camps, women's labour camps or children of mothers deported for forced labour; abandoned children; children to whom special orders applied; children sent to Germany for forced labour.

The phrases 'good impression' or 'bad impression' which played such an important part in the terminology of the race experts who selected girls for reproduction on the Führer's behalf were used here too, and the difference they made was that between life and death. Thus in the files of the 'L office' there are black lists drawn up by the Reich Adoption Office of non-valuable children, and of 'Germanic' and Austrian, Polish, Russian and Rumanian children. These lists date from the end of 1941, and also include the names of 'non-valuable' adults. They can have been drawn up only for submission to another agency, the agency entrusted with 'taking care' of undesirables. By what mysterious channels did these lists of 'non-valuable' persons reach the officials of the Reich Adoption Service, passing through the net of the medical examiners' sifting process? This department certainly took part in kidnapping too, but the explanation of the administrative anomaly must be sought in the way in which certain kidnappings took place. In Polish and later in Russian villages the kidnappers conducted large-scale 'search operations for children of Germanic origin'. The children were rapidly assembled in a square or school playground and, after a summary selection, were taken by truck to Poznan or Lublin, where huge selection camps operated. There only 10 per cent turned out to be valuable, and it was more convenient to send the remainder to a neighbouring camp, or put them in a train leaving for Auschwitz, than return them to their homes.

Before being simply taken from their parents without further formality, children were spotted, sometimes several days in advance, by the physiognomists of the Sipo (security police). Children fled at the approach of an open Volkswagen prowling the streets of towns and villages, so, to avoid frightening them too much, the 'Brown Sisters' were were called in to help. These wore brown uniforms: a female counterpart of the SA and Nazi Party uniform consisting of a wide, dark brown dress shaped like a sack of potatoes with white collar and cuffs and a grey apron. 'Blonde Sisters' were employed to reproduce the species for the Führer, and 'Brown Sisters' to skim the conquered territories for fair-haired children. SS terminology was not very imaginative, so the word 'nurse' should be avoided in describing these women specialists in kidnapping, denunciation and cultural genocide. In Poland they were and still are called 'The Brown Sisters of the SS'.

Actually these women belonged to the NSV, established in 1933 to devote itself to the welfare of the German people. Under the aegis of the Youth Office, their activities covered the whole of Europe; they established maternity homes and crèches for the armies of occupation and reception centres for children 'captured' from the enemy. To those who suffered under them, these fanatical Nazi women, totally dedicated to the Führer, were perhaps even more loathsome than the killers of the SS or the SD; stony-hearted robots was one description. The sight of one of these women – often of mature age – brutally snatching from its mother's arms a baby who was smiling at her remains an intolerable memory to those who experienced it.

The special training of the 'Brown Sisters' included intensive courses in which they were taught the racial criteria by which Nordics could infallibly be distinguished, and they were instructed in how to observe a child without being noticed themselves; they were also taught ways of abducting it in the street, at home or at school. These courses were being organized by a special department of the RuSHA or the Gestapo in Berlin even before the outbreak of the war. Later, in view of the poor results obtained by improvised baby raids, the selection and training of the physiognomists became stricter. Students of the 'Physiognomic Brigade' ended by being able to spot at a glance the fault that might be concealed beneath a shock of fair hair or in blue eyes that were too Slavonic.

Their technique of approaching children in the street did not vary greatly. A hungry child would be offered biscuits, sweets, sometimes even a bar of chocolate or a slice of bread, thus creating an opportunity to question it about its parents, its home, the colour of its brothers' and sisters' hair. That same evening they submitted their list of names and addresses to special teams of kidnappers, with carbon copy to the RuSHA. The latter would carry out a rough preliminary sifting by consulting the records at the town hall. Several days would elapse, and then the child would be taken, the abduction generally taking place at night. The child's parents would never see it again.

Now it was the turn of the medical and other examiners. Children who passed the tests were taken to a Lebensborn reception centre; the others generally disappeared without trace, often being dispatched to a concentration camp. Luckier children might be returned to their parents without explanation. The kidnapping game does not seem to have been played in accordance with any fixed rules. The decision whether a child was to be sent to its death or back to its parents depended on the whim of a medical examiner or even of the SS man on guard at the door.

The worst fate was reserved for 'Nordic' families refractory to Nazism. Not content with taking away their children, the SS vented their fury on the parents, who often suffered the death penalty as a result. Hundreds of families were thus liquidated. In such cases the Lebensborn and the NSV shared the booty; the former took the children under six, and the latter took those between six and twelve and placed them immediately in such special institutions as boarding schools, Reich schools, Napolas (Nazi political schools) or the BDM.

Psychological methods were used to make a child forget or even hate its parents. He would be told they were dead, and that there was nothing honourable about the way they died. The mother would be said to have been of doubtful morality and to have died of tuberculosis, drink or other shameful disease, while the father had died of cancer or drink, or been killed by Polish bandits. The object was to give the child a sense of inferiority about its origins and of gratitude to the Germans who had rescued it from the degeneracy of its home environment.

In the German Federal Republic we met a young woman who, at the age of five, had been taken to a church by the Germans and shown a bishop's coffin and told it was her mother's. Some years later the child was traced, but she refused to go back to her mother, who had survived deportation. 'I had stood by my mother's coffin once,' she said, 'and I did not want to do that again.'

The faces of the 'Brown Sisters' can still be seen smiling at their unsuspecting victims on old photographs faded by time. How many lives did they wreck in Poland alone? Two hundred thousand? Perhaps more? The number will never be known.

After the war the International Refugee Organization searched for these children in the Allied occupation zones. Only between 15 and 20 per cent were traced, often thanks only to a typical piece of Nazi officialdom. Industrious, orderly and unimaginative, when they contented themselves with a literal translation into German of Polish names they greatly facilitated the work of the Allied investigators.

15. Valuable or Not?

'When I think that with those blue eyes and fair hair they talk Polish, it seems absolutely incredible.'

HANS FRANK, Diary, 30 May 1940

All Polish children without exception had to be submitted to racial examination by the RuSHA. Those thought fit to serve the Führer or considered suitable for Germanization had to be taken on a fixed date to the Youth Office, where they were immediately separated from their mothers, who were unceremoniously told that there would be no point in trying to see them again.

Thousands of children between the ages of six months and twelve years were thus assembled, notably in the SS children's home at Kalisz, which was in fact the principal Lebensborn home in Poland. The children stayed there for about six weeks before being sent under a false identity to a Lebensborn home in Germany or directly to German families awaiting their turn on a long list of would-be adopters. It was at Kalisz, near Lodz, that the first phase of the Germanization programme took place; here the names were changed and new places and dates of birth allotted. This transit centre opened in 1940.

Zygmunt Krajewski, born in Poznan on 17 April 1933, was one of the few who succeeded in escaping from a Lebensborn home, that at Gmunden, in Austria. This is his story:

'I was taken by force from my family on 20 May 1943, and sent to the camp at Kalisz. I stayed there for four months and had to learn the Hitler doctrines. Then I was taken with many other boys to Gmunden on the Danube. [This was the Oberweiss home, the head of which was Kurt Heinze, who collected the children from Poland himself.] Here the food was satisfactory and life was tolerable.

'Hell began the day we were forbidden to talk Polish. We were accommodated in groups of about forty boys, aged between six and thirteen, in a big hall. The daily routine was as follows: We got up at six o'clock, did gymnastics, then washed and made our beds, and had breakfast at 7.30. Then there was roll-call, and we started work in the garden, digging, watering and so on. The work was very hard, and it was not unusual for a boy to fall ill through exhaustion. We were punished for the smallest mistake or act of disobedience by being beaten, or not being allowed lunch, or having to work for an extra hour. I could not reconcile myself to denying my nationality, so I went on talking Polish. For this I was often tied to a post and beaten but, as I was strong and refused to give in, I managed to stand it. In winter it was worse, because we were lightly clothed in the Hitler Youth fashion. Germans often came to the camp and picked out boys they liked. The child would be told his parents were dead and that he was going to get new ones. In any case, our surnames and forenames had been Germanized. I still remember precisely what happened while I was there. When the Germans began talking to me and offering me sweets, I knew at once what they were at and answered in Polish. Of course the resulting punishment was dreadful, but I preferred it to disgracing myself and going to a Hitler family. They had no success with me. Eventually, I managed to run away and managed to get home with the aid of some good people, but I still suffer from a bladder complaint because I froze so during my flight.'

Zygmunt Krajewski's feat in escaping from Gmunden in Austria and finding his way back home to Poznan on foot was unique. Most such attempts failed, and those who made them were severely punished. A child, in the situation of an escaping prisoner-of-war, had against him lack of physical endurance, fear of the police dogs, and lack of a sense of direction. Some young escapers went in circles round the home, hiding in the trees during the day and trying to find something to eat at night before being found asleep by a patrol two or three days later.

No doubt it was because of its high surrounding wall that the monastery at Kalisz was chosen by the Lebensborn leaders; it seemed to preclude all possibility of escape, as well as any attempt by parents to recover their children. Nevertheless, in spite of the sub-machine gun of the guard at the main gate and the dogs that roamed the park at night, some got away with the aid of a man known to the children as 'Papa Stanislaw'. We went to see Stanislaw Kuczynski at Kalisz. He told us:

'I was a kind of general handy-man at the monastery. I repaired windows and doors, lugged milk churns and baggage, and helped unload the provision lorry and cart away the rubbish. Children often came to see me in my carpenter's workshop, but I particularly remember a little fair girl named Christina, who looked so desolate and sad at being so suddenly deprived of her family. When I muttered a few words of Polish to her, she came over to me and was overjoyed. I secretly slipped her some sweets and home-made toys which she distributed to her friends.

'One day a Poznan woman spoke to me in the street. As a result of patience, astuteness and courage, she had managed to trace the child whom the Germans had taken from her a few months before. She was Christina's mother. She told me she had made up her mind at all costs to get her child out of the hands of the SS. I told her to be very careful, and to wait until I found a way of getting the little girl out. It wasn't easy, because the German teachers were strict and let no one in or out unchecked. At last we found a way, on the day when the Germans wanted to have a load of waste paper dumped at the public refuse tip outside. The carter and I hid the little girl under the paper. I arranged for the mother to be waiting near the tip, and she and her daughter fled to a hiding-place in the Poznan region. That was the first escape from Kalisz. There were others later, but I don't know what happened after 1943, when I had to get away myself, because things were beginning to get too hot for me.'

Stanislaw then showed us the chapel at the entrance to the park.

'It was there that the Germans shut children who refused Germanization. They had to stay there in the dark on their knees with their arms crossed for hours. They wept, and soon fainted. They were punished like that for saying something in Polish or talking about their parents. They were beaten and deprived of food. But even apart from that, the children were always sad. They lived in fear and were homesick, and the German supervisors felt nothing but hatred for them, because they were nothing but little "Polacks" and did not belong to them.'

We asked how many children he saw while he was at Kalisz.

'Thousands. They arrived in parties of sixty or eighty. The healthy ones stayed here for only a short time; they were sent on to Germany. Only the weakest were kept in the monastery. Others were coming and going all the time. The Nazi education programme lasted several weeks. Discipline was severe, very severe. There were actually deaths among the children.'

Not far from the monastery is a cemetery in which most of the graves are of 'victims of Nazi barbarism'. Tadeus Martyn, a member of the Polish Commission for Hitlerite Crimes, led us to a small white grave covered with flowers and there told us about young Zygmunt Swiatlowski:

'He was taken from his parents against their will at Poznan and brought here, to the SS home at Kalisz. He felt himself to be Polish, would not be Germanized, and refused to speak a single word of German. He was often beaten by the staff. One day, after refusing to greet a German in German, he was killed on the spot by the woman in charge of the institution, Johanna Sander. The children who died in the home were buried anonymously, but the German who buried

Zygmunt revealed his name to the Polish woman caretaker of the cemetery. So this grave remains the only memorial to the martyrdom of Polish children at Kalisz.'

We read in a Polish publication, *Guide to the Second World War 1939–1945*, that many children aged between one and fourteen years, deported from various regions and interned in the camp at Potulice (district of Gdansk, formerly Danzig) for the purpose of being racially examined, died there as the result of maltreatment. About five thousand are buried in a special cemetery.

A large number of girls, bureaucratically kidnapped by merely being summoned to an office by some SS official at Poznan, Lodz or elsewhere, were sent by order of the Lebensborn organization, officially in charge of them, to the Illenau school at Achern in Baden, an SS barracks now used by the French Air Force. The figures show that girls were often given priority over boys, and this applied both to the 'Germanic' provinces of Poland and to the rest of the country, as well as to Czechoslovakia and Yugoslavia. We spoke to some of the now grown-up victims who had the rare fortune to be repatriated after the war, and we also interrogated German families to whom some of these girls were entrusted temporarily (that was, not for adoption) by the Lebensborn organization. Alicja Sosinka, born at Lodz in 1935, taken from her mother in September 1942 and turned by Dr Tesch into Alice Sosinger, born at Poznan, told us:

'My mother was ordered to take me to the Youth Office in Copernicus Street, where the orphanage is now. I was separated from her the same day and sent to Kalisz a few days later. On 30 October we were transferred to the boarding school at Illenau with a party of other Polish girls. Life in that barracks was a real nightmare. We were branded on the left hand and the back of the neck [the scars are still visible]. We were very frightened by this branding, but in fact it was not painful. One day we were told: "You will give birth to two or three Germans of good race and then you will disappear."

'We were also continually given injections. I now believe they were hormonal injections, intended to make us reach the age of puberty more quickly. The other girls, Poles, Czechs, Hungarians and, I think, Rumanians, were subjected to the same treatment. Every now and then the SS put us through another racial examination, each more severe than the last. The girls who did not pass were never seen again.'

We asked Alicja what happened to girls who refused Germanization and such treatment.

'There was no such thing. We were much too frightened. After the war, I did not want to go back to Poland at first, for after Kalisz or Illenau the mere idea of being put in a camp again made me ill . . . and for months, when my mother came to tuck me in at night I used to jump out of bed and stand to attention. . . I prefer not to talk about it any more.'

For a few months Alicja was placed with a peasant family in the Illenau neighbourhood. Their daughter told us that her parents had very much wanted to adopt her, but unfortunately that had been impossible, because the SS had intended to take her back for reproductive purposes when she reached the age of fifteen or sixteen.

Frau Annemarie Zink, in whose family the girl Danuta Wutzow was brought up, told a similar story. She was the daughter of a prominent Nazi, had been a member of the BDM, and had moved heaven and earth to have the child adopted, because the whole family was devoted to her, but in vain. When the Lebensborn people had handed her over at the outset they had said: 'We will fetch her again as soon as she is eighteen.' When we asked why, Frau Zink explained: 'Those children were selected in advance to be the mothers of children of good race, so no one was allowed to adopt them.'

Urzula Nadolna, of Rogozno, was aged eleven when Kurt Heinze and Günther Tesch turned her into Ursula Nadler, an inmate of the Oberweiss home run by the Lebensborn organization in Austria. She too was not adopted by the

Austrian foster-parents to whom she was handed over as a *Polish* child. 'I was told I was too big to be adopted, and also that I would soon be able to have children myself,' Urzula said. Other witnesses, both Polish and German, mentioned the SS plans for these girls who were 'recovered' in Eastern Europe solely for the purpose of reproduction later on.

Such projects are not mentioned in any surviving document, for a few weeks before the collapse of Germany the SS destroyed the files concerning kidnapped children along with those concerning children who were the result of planned reproduction. No one in any position of authority in the Lebensborn organization has ever admitted that such schemes existed. It would have been surprising if they had, for the above statements throw new light on the purposes behind the SS kidnappings. They clearly show that while very young children with no precise memories were ideal raw material for Germanization, the fate of the older ones was infinitely more hazardous.

Boys with whom the Nazi indoctrination methods succeeded were to swell the ranks of the SS mercenaries needed to control the slave populations. As for the girls, having been made precociously fertile, they were intended to be used in what were popularly known as the human stud-farms. What the Lebensborn organization perhaps did not dare do to German girls could have been done with impunity to racially valuable foreign females, who could be kept alive or liquidated in accordance with requirements. The selection of these 'breeding women' for the thousand-year Reich would have taken place automatically, in the Lebensborn homes. According to the results, little Alicja from Lodz, brought to puberty precociously, would have had the right to one, two or three confinements in an SS maternity home. Carefully selected men would have mated with her out of personal pleasure and duty to their country. Then one day some Dr Ebner on duty would have given her an injection and Alicja would have ceased to exist. Thus all trace of foreign blood used for the Nordic repopulation of Germany would have been obliterated for ever.

So much for the systematic, bureaucratic kidnapping. What of the wild or random kind? We have seen how the 'Brown Sisters' and other SS specialists in physiognomy worked in the towns and villages of Poland. By the end of the war 200000 or more children had been kidnapped.

In spite of Russian military pressure, in 1942, 1943 and even 1944 orders for the recovery of Nordic blood rained down on the desks of the special units engaged in the work, and kidnappings grew steadily more numerous. In the street, at school, at home, at kindergartens and even in public parks children were the victims of raids which nobody dared oppose. A climate of terror prevailed. One day orphanages would be emptied, next day children born of German fathers and Polish mothers would be seized – rarely the reverse – and a few days later there would be another round-up of children in reprisal for a partisan operation. Children who made a bad impression left for forced labour or for Auschwitz; those who made a good impression would be presented at Kalisz to Frau Inge Viermetz, who paid several visits to Poland. So did the bald-headed Dr Ebner, who was 'astonished at so much fair hair', and Kurt Heinze, the indefatigable head of the Oberweiss home, who continually shuttled between Kalisz and Germany. He escorted whole train-loads of children whom the Lebensborn organization rapidly placed in State schools or families. Sometimes official and 'wild' kidnapping went hand in hand, mingling with or rather supplementing each other, as was illustrated one night in September 1943 at the village of Rogozno in the Poznan district, in the province the Germans called the Warthegau.

Before the war roughly a third of Rogozno's inhabitants were Poles, a third were ethnic Germans, and a third Jews. By 1943 no Jews remained. The *Volksdeutsche*, the ethnic Germans, wearing the Nazi arm-band, made life hard for the Poles who worked for them. One of these was the mayor, who did not need much prodding before summoning to the town hall the fair, blue-eyed children picked out by the race office scouts. A first sifting of 150 children led to the selection of eight 'valuable' specimens. Among them were Urzula (found in the Russian zone in 1945 and promptly repatriated); Kasimir (who also returned in a Soviet army lorry); and Eugenia, whom we shall meet again at Flensburg in Germany, where she started a new life as a German. There were also two cousins, Leon and Alojzy Twardecki, who were kidnapped the same night, but who had a very different destiny.

Leon and Alojzy, like all the other children born of unmarried mothers, were examined by the RuSHA. The mothers were told it was a medical examination in the child's interests. A fortnight later, on 23 September, the round-up began. Some children were summoned to the town hall, while others were roused from their sleep at night by three SS men led by the mayor (who now lives in Berlin). One Rogozno woman, however, hanged her child from a balcony rather than let the Germans take him. But Leon and Alojzy were among those dragged off to the station, where a train had been waiting since the previous day to leave for Poznan, from where the children, with others, were to be sent to Kalisz. Alojzy's mother, whom we met at Rogozno, remembered that night:

'I had one son. It was at three a.m. on 23 September 1943 when the Hitlerites took him away. I was in complete despair. They came with dogs and made Alojzy get out of bed. Then they left. We mothers were ordered not to follow them, but we did. All the way to the station. On the way I managed to tell my boy to hide behind a bush or a tree or somewhere, but he refused, and said: "But, Mother, they'll beat us, I prefer to go. But I shall come back. There will certainly be toys there too." '

Leon, Alojzy's cousin, was eleven. He was small and slightly built, but his hair was fair enough. The RuSHA suspected that both boys had been fathered by village lads of German stock. How could it be otherwise with boys of such obviously Nordic type?

At Kalisz the Germans gave the two cousins the same

surname, Hartmann, literally 'hard' or 'tough' man, a translation of their Polish name. This enabled Leon after his return to Rogozno to remember the name of his cousin, then aged four, with the result that the latter was found at Koblenz, where he was living as the son of a Herr Theo Bindenberger, a loyal member of the Nazi Party. (The story of his eventual repatriation is told on pages 221–6.) Leon, on the other hand, had lived up to his name and refused to be Germanized. He had survived the discipline in the camps of the Hitler Youth into which he was forcibly enlisted, and after the liberation found his way back alone to his village. Today he refuses to speak a word of German, and above all refuses to forget Kalisz, Oberweiss and Heinze.

'Children who refused to take part in the farce received terrible corporal punishment. If we spoke Polish we were savagely beaten. Another punishment was to be shut in a small chapel, where we had to kneel for hours on the cold stone floor and were given nothing to eat. In the Hitler Youth camp near Vienna there were two Poles to every ten Germans or Austrians, who treated us like lepers. I was never adopted. I was too big and too Polish, and no one wanted me.'

Eugenia Ewertowska, aged eight, renamed Irene Ewert by Dr Tesch's department, was suspected of being the natural daughter of an ethnic German inhabitant of Rogozno. She is blonde and Nordic in type, but, unlike Alycia or Danuta, who were sent to Illenau, she was considered to be of German descent, so she could be adopted. Her mother was one of the weeping women who followed the children to the station that night. Sometimes a mother would try to grab her child and drag it behind a tree, but the trained Alsatians growled and order was quickly restored. Eventually the station was reached. The children were herded into the train to the accompaniment of their mothers' cries and disappeared behind the drawn curtains of the compartments. The mayor raised his hand, the train was just about to move off and the SS with their dogs were just preparing to leave when their attention was suddenly attracted by shouting from the front of the train. The driver was gesticulating wildly and pointing to the rails, on which the mothers had laid down less than three yards from the front of the engine. But dogs and rifle-butts were put to work, the line was quickly cleared, and a few minutes later the train was steaming through the sleeping Polish countryside on the way to the town of Kalisz. Irene was never to see her mother again – and has no desire to do so. Her case, to which we shall return later (pages 241–3), was cited at the Lebensborn trial as an instance of kidnapping, Germanization and adoption.

Urzula Nadolna was also in the train, and still does not know why she was not taken away in the middle of the night like Aloyzj, Leo and Eugenia.

'The Germans had summoned us to the town hall with our mothers a fortnight earlier. They wrote down our names, some on white cards and others on red. At least 150 children were examined that day. On the red cards they entered the names of children who had blue eyes and fair hair. There were eight of us from Rogozno altogether, two girls – little Eugenia Ewertowska and myself – and six boys. So my mother took me to the town hall, and it was from there that we were taken to Kalisz. The Germans said they were taking us there for health reasons, but actually it was a transit camp where they collected children in the name of the German race.'

The great majority of the ethnic Germans in Poland, Czechoslovakia and Yugoslavia loyally followed Hitler, and their collaboration even extended to the kidnapping of children of people among whom they had lived for centuries. After their overnight transformation into members of the master-race, they behaved, especially in Poland, in a way that makes their post-war expulsion easy to understand.

In the files recovered by the Polish Red Cross a considerable number of these ethnic Germans appear under the heading of adoptive parents of selected children. Being far better off than their Polish neighbours, to say nothing of the Jews, whose property was expropriated for their benefit, they followed the trend of the times towards having large families, and by having children of their own or adopting kidnapped children they faithfully followed the racial principles of the RuSHA. On the principle of first come, first served, they were often to be found in the Youth Offices whose sole purpose was selecting other people's children. Later, when they were expelled, they made off with their adopted children, thus creating new dramas. Here is one among many.

Jan Chrzanowski, born on 5 April 1940, was subjected to a first examination at Lodz when he was barely a year old, and his mother was sent to a labour camp in Germany. As he presented some advantageous Nordic traits, he was again examined in March 1943. Dr Grohmann, of the race office, reported as follows:

> According to his physical appearance, he is a harmonious blend of Nordic and East Baltic types. There are no racial objections to his being cared for by a German family... Frau Martz, living here at 11 Ulrich von Huttenstrasse, is willing to accept him... Frau Martz requests help in having the child's name changed immediately. She objects to having to register him everywhere under his Polish name.

Seventeen years later Jan, who had long since become Johann, allegedly born on 5 April 1940 at Gotha, was found by pure chance by the Polish Red Cross. Meanwhile he had been Germanized and had accompanied his ethnic German family to Germany. He refused to return to Poland on the ground that he had never known his mother.

How lightly the Germans regarded taking Polish children from their parents is shown by the adoption documents. To give an example: 'The child was taken from its father, a carter, when he was arrested while trying to enter the ghetto.' This happened at Lodz. The adoptive parents were an ethnic German childless couple named Kleiber, who pestered the authorities until the child was theirs. The only other comment in the documents on this case of a child abducted because his father committed the crime of trying to enter the ghetto was that 'the adoption costs amounted to thirty-seven marks'.

In the case of the infant Ryszard Jaskulski, Dr Grohmann made a remarkable discovery. 'Racial diagnosis is difficult in the case of a very young infant,' he pointed out, and went on to remark that at this stage 'many characteristics have not yet appeared and are still in a state of more or less untypical development'. Nevertheless, in view of the fact that the child's mother, Wladyslawa Jaskulska, was suspected of having had relations with the Jew Abraham Kohn, the head of the Youth Office medical department concluded that little Ryszard (he was only thirty months old) was a first-degree Jewish half-breed. 'He is very small for his age and very under-developed. Jewish influence is therefore to be regarded as probable, and I suggest that this child be transferred to the ghetto,' he wrote in a letter of 12 February 1941. The Youth Office promptly informed the Gestapo of the transfer, and a half-sheet of paper signed by the senior member of the Jewis community at Lodz confirmed the child's reception into the ghetto. Two days later the Gestapo scribbled on a sheet of paper a note to the effect that in view of this no more maintenance allowance was to be paid. After the war no one came forward to claim the child. It would have been pointless in any case.

When the heads of the Lebensborn organization were tried at Nuremberg in 1948, a defence witness, Maria Martha Heinze-Wisswede, claimed that the Lebensborn organization had been favourably inclined towards the Jews. In other matters this former 'little blonde sister' did not have very clear recollections, but she remembered a 'Jewish child' received into a home as if it had been her own son.

PRESIDENT OF THE COURT: Are you aware of a case in which a Jewish child was accepted into a Lebensborn home by mistake?

WITNESS: Yes, I remember a case.

PRESIDENT: Please tell us about it.

WITNESS: Siegmund Raschke was about fourteen. He was at Oberweiss when I was there. Then we found him a place as an apprentice.

PRESIDENT: Was it noted in his papers that he was of Jewish origin?

WITNESS: Yes. We received the papers from the Youth Office concerned, and they stated that Siegmund Raschke was Jewish.

So the Lebensborn people had apparently saved a Jewish life. And in this connection it is worth noting Ebner's ambiguous claim under cross-examination at Nuremburg. To the question 'You would have accepted Jewish mothers in your homes?' he replied: 'Surprising though it may seem, yes, that would have been possible. . . . I remember that a Japanese woman applied to be confined in a Lebensborn, and was admitted. After all, we did no propaganda in Jewish circles, that was why no Jewish mother ever applied to us.' A racial error of such magnitude as the case of Siegmund Raschke seemed worth following up, but all efforts were vain. Leo Twardecki, Urzula Nadolna and many others who were at Oberweiss for a long time were unable to remember this boy who, in spite of his Nordic appearance, had Jewish blood in his veins. Moreover, they thought that fourteen, the age attributed to him by the 'Little Blonde Sister', was not very likely, as the oldest child there had been under thirteen.

When we tracked down Maria Martha Heinze-Wisswede at Neustadt-am-Rübenberge, near Hanover, she refused to answer our questions, claiming she had never heard of the Lebensborns, though she was the only person who might have been able to give us information about this alleged rescue of a Jew by the most racist organization in Nazi Germany. As for her husband Kurt Heinze, who had also displayed a marvellous memory at the Nuremberg trial when it came to posthumously establishing the German parentage of Polish children claimed by their real families, he had suddenly 'gone away'.

Jan Woszczyki is one of those Poles who have undertaken the task of passing on to future generations a true picture of their fellow-countrymen's sufferings during the years of war and occupation. His account is free of hate, and merely gives the facts about what millions of Poles, including himself, lived through. Since 1945 he has been touring Poland, lecturing in schools with a worn album of fading photographs, telling the present generation of children what life was like in the child slave labour camp at Lodz. Jan Woszczyki himself was not considered 'racially valuable'.

'It was on 2 February 1941. I was going by train from my home village of Porabka, near Sosnowiec, to Kluezborg, a village in the same district, near Katowice. After a few kilometres the train suddenly stopped, and Gestapo men jumped in. They drove us out with rifle-butts, and those who were not quick enough were attacked by dogs. A few hours later we were in the prison of Lublin.

'I arrived at the camp at Lodz on 27 September 1942, after spending eighteen months in various prisons. I was twelve then, but when they arrested me I pretended to be two years younger, hoping they would let me go. The camp was not yet ready to receive prisoners; it was just being organized. I was given the number 127. Soon after our arrival we were put to work levelling the ground, marking out sites for huts, putting up barbed wire, erecting electricity standards, installing electricity, and digging machine-gun emplacements. The camp was intended to hold three or four thousand children.

'After a month I had to appear before the race committee. After the examination I was classified as not Germanizable. My appearance did not fit in with their requirements.

'More parties of children arrived every day. Those classified as "valuable" were put in separate huts. They were fairly well treated, at any rate in comparison with us. They came from all over the place. There were Russian, Czech, Belgian, and French children, and actually a few blacks [born of German mothers and African fathers in the Rhineland, occupied up to 1936 by the French, who had a large number of colonial regiments there, chiefly Senegalese], many Germans and a few Jews, but above all a huge number of Poles. The half-litre of soup that was practically all we were given to eat all day contained some strange chemical poison that caused burning in the stomach and intestines and caused kidney trouble. We swelled everywhere because of under-nourishment, which also affected the nerves, on top of which there was the cold, the blows, and the exhausting labour. Many went out of their mind. Children who wetted their beds were sent to Block 8, which had no doors or windows. The children were given blankets that were as thin as spiders' webs, and the temperature fell to minus twenty degrees in winter. During the night they froze. Next morning we had to use picks to cut the stiff bodies away from the plank beds, load them on to carts and take them to the Jewish cemetery adjoining the camp. We flung them into a mass grave, threw lime over them and covered them with earth. Sometimes they were not quite dead. When they began to suffocate through lack of air, the earth over the grave moved like a cornfield in the wind. As soon as they were suffocated the movement stopped and the earth was still again. On an average 120 of the 3000 or 4000 children died every day. They died of cold, blows, hanging or shooting. We who survived know that they went through the whole gamut of possible cruelty.'

In the early stages of the occupation Dr Witaszek of Poznan organized a resistance group. His knowledge of chemistry enabled him and his assistants to concoct a slow-working fatal poison. Men and women of his group secretly distributed small bottles of the lethal mixture to canteens and restaurants used by the occupation troops. With the help of Polish cooks, everything went well for a few months, until they were denounced.

The doctor was arrested on 25 April 1942, and was beheaded in the fortress of Poznan with three of his comrades. His wife was sent to Auschwitz a week later. Two of their daughters, aged four and six, were taken away by the Sipo, and were sent by way of Kalisz to the Lebensborn home at Bad Polzin. One was adopted by a German family in Mecklenburg and the other by an Austrian family. Thus the children of racially valuable 'bandits' were treated just like others.

Mrs Witaszek survived Auschwitz and, in spite of immense difficulties, succeeded in tracking down her daughters, even though Dr Dükker, the head of the Bad Polzin home, had, like others, taken the precaution of destroying the papers concerning them. At Poznan Mrs Witaszek told us:

'Years afterwards my younger daughter told me she had often been kept awake at night, wondering why I had sold her to a foreign family. Did I have so little money that I had to sell her? Children at that age were simply incapable of understanding what had happened to them.'

A few years ago a ceremony was held in Wolnosci Square at Poznan to commemorate the memory of Dr Witaszek and his comrades, whose heads now rest in the heroes' cemetery there. The German authorities wanted to have the brains of a Polish scientist examined, so the heads had been sent in jute sacks to a research institution in Germany, where the professor in charge had the idea of preserving them in glass, so that brains of the Polish intelligentsia might be studied. The bodies were never found.

Children of 'bandits', children picked up by chance, children arrested by special orders and declared to be 'valuable' before being sent to camps, children rounded up in huge raids under the code names 'Operation Hay' or 'Operation Gypsy Baron' – the list is long, and the fate of the victims hard to discover. Parents did everything possible to trace children who were unaware of their existence and will never know the distress their absence still causes. In some Polish villages the grief is still so vivid after thirty years that one ends by wondering how such a thing can be possible. For child kidnappings, deaths, arrests, internment were succeeded by consequences of a different order, both moral and physical, due to illness, absence, memories.

Sonia Gruszka has always refused to talk about her concentration camp experiences, even to her daughter, but agreed to speak to us because 'the Frenchwomen at Ravensbrück had been kind to her and all the other detainees'. (We found similar evidence of gratitude to French deportees throughout our travels in Poland.) Mrs Gruszka was twenty-two when she was picked up in a big raid and became a prisoner of the Gestapo at Lvov. She was deported to Germany, and with a party of other Polish women was put to work on the land for a rich farmer in Württemberg. By a remarkable chance her husband, a Polish prisoner-or-war, worked at the same farm. They sometimes managed to meet surreptitiously at night in the orchard, but after a row with the farmer he was sent to Dachau and she to Ravensbrück.

'When I arrived at the camp I was five months pregnant. Many women had babies in the camp, but I was told that only fair-haired children had the right to live. The others were either drowned before their mothers' eyes or dumped on the refuse heap. I prayed that my child would be born with fair hair and blue eyes, and that is what happened; my daughter Lucyna was allowed to live. I was able to keep her for a month, but our forewoman at the farm had already applied for her. One morning the SS came and my daughter had gone. After the liberation the German woman refused to return her to me, but Polish-speaking American soldiers helped me.'

Lucyna's Polish identity card describes her as having been born at Buchenwald, though it should have been Ravensbrück, the women's camp; Buchenwald was a man's camp only. The mistake presumably arose because both camps came under the same administration. When we met her she said:

'For years the fact of having been born in a camp made me think I was different from other people. So few of us survived. That made me feel ashamed, and I never dared show my identity card.'

Lucyna Gruszka's father died at Dachau. She survived because at the age of four weeks she made a good impression on the race experts.

Would she have been Germanized? Would she later have

The last photograph of Eugenia Ewertowska before her abduction in 1943.

Thirty years later her mother is still waiting for her.

By 1980 the Greater Reich was to be populated by 120 million Nordic Germans.

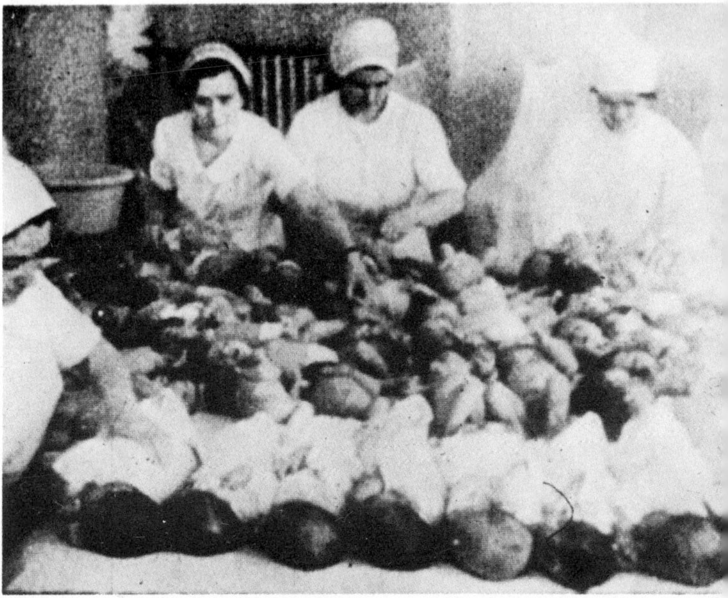

Babies left to their fate in the Lebensborn maternity home at Steinhöring after the leaders of th
organization had fled.

Above
More than 200,000 Polish children were taken away from their parents.

Children were rounded up and taken in truck-loads to camps where a first 'sifting' took place.

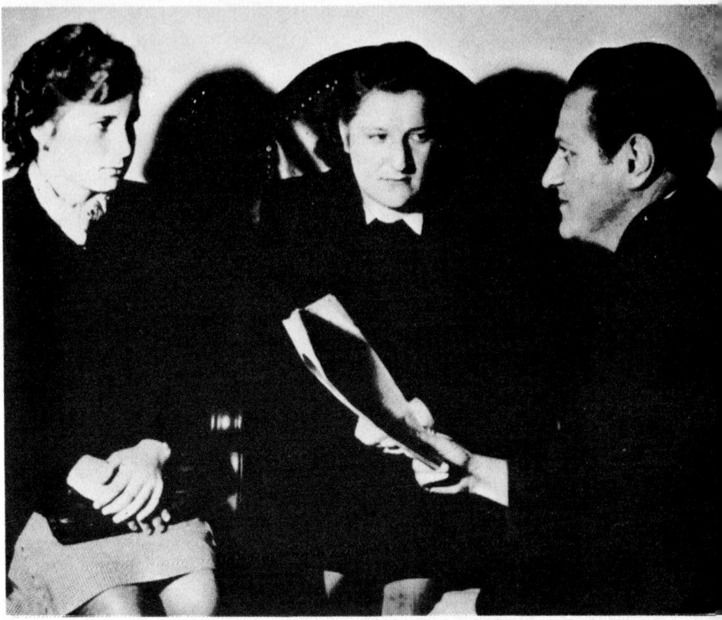

Maria Hanfova and Maria Dolesalova, survivors of Lidice, giving evidence at the Nurembourg trial.

Parents who refused Germanization were sent to concentration camps. The race experts kept only 'valuable' children.

Yugoslav 'bandits' babies arrive in Austria in 1943.

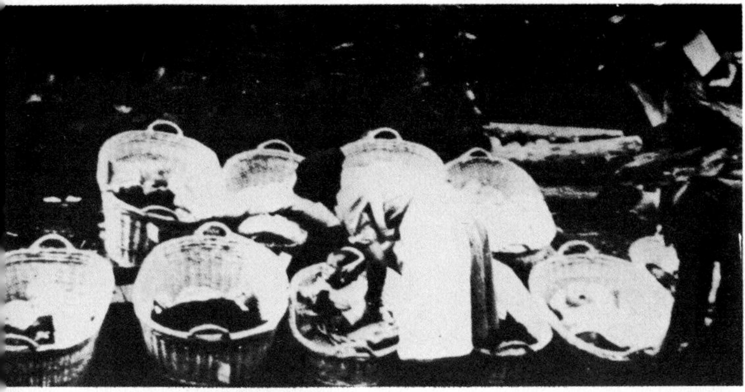

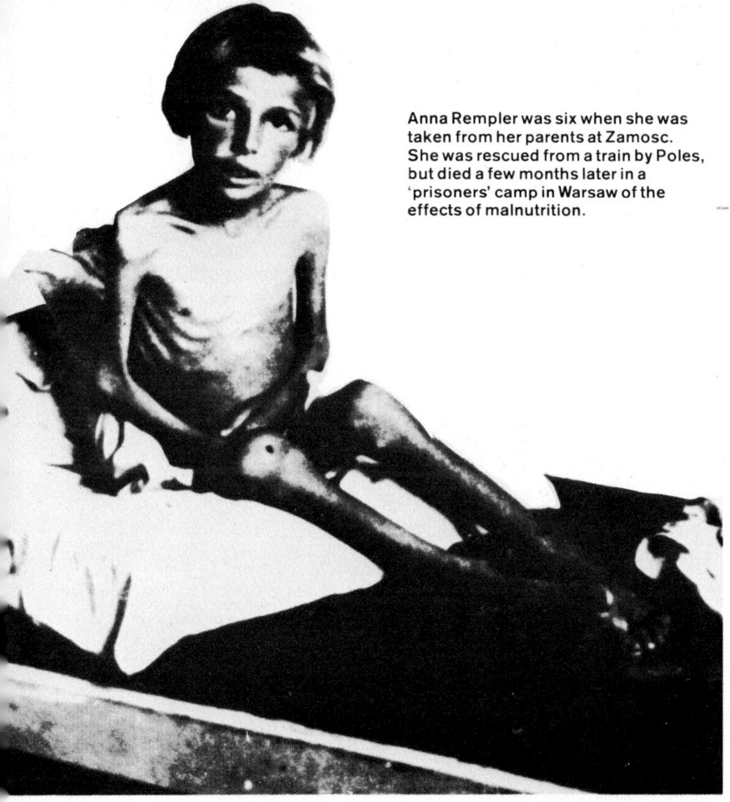

Anna Rempler was six when she was taken from her parents at Zamosc. She was rescued from a train by Poles, but died a few months later in a 'prisoners' camp in Warsaw of the effects of malnutrition.

Above
'Non-valuable' children were destined for forced labour. The camp at Lodz in Poland.

Left
Racial selection took place in the first instance on the basis of the child's appearance.

Fewer than 20 per cent of the children were found after the war.

been branded on the left arm and the nape of the neck to serve the cause of the Greater German Reich? How many babies born in concentration camps, labour camps and other places where slaves were assembled were killed for the simple reason that their eyes were the wrong colour? The Polish Red Cross estimate is more than 40000, to which must be added an incalculable number of enforced abortions ordered in the name of the purity of German blood.

16. The Forgotten Trains

> 'The children must be educated. They must be taught obedience, industry, total subordination and loyalty to their German masters. They should be able to count up to a hundred, recognize badges of rank, and be trained for jobs such as farm-worker, kitchen hand, locksmith, stone-cutter. The girls should learn land-work, weaving, spinning, dress-making, etc.'
>
> Reichsführer's circular, 18 June 1940

Odilo Globocnik, who would commit suicide after the German collapse, had two passions in life: killing Jews and archaeology. In his Gruppenführer's uniform he was a well-known figure in Lublin, and he was badly in need of an idea that would please his friend Himmler, to whom he owed so much. It was Himmler who had accepted him into the SS when he fled to Germany after murdering a Jewish jeweller in Vienna in 1933, and who, six years later, when Globocnik had become Gauleiter of Vienna, hushed up a huge foreign-exchange swindle in which he was implicated. It was also to Himmler that he owed his promotion to the SS rank corresponding to that of general and his appointment to the task of 'disinfecting' the Eastern territories that the German armies had overrun.

In the Lublin area Globocnik thought he had found traces of ancient settlements once occupied by Germanic tribes. Obeying the call of his more-or-less Nordic blood, the ex-convict set with alacrity about the task of restoring the territory over which he ruled to its 'real' owners. It was a large and over-populated area, not least with SS troops and officials, for Himmler had set up his Polish headquarters there. The towns had names such as Lemberg, Shitomir, Lublin, Winnica, and there were smaller ones such as Tomaszow, Hrubieszow and Zamosc. In the autumn of 1941 Himmler agreed to the Zamosc area being made the first big area of German settlement in the heart of Poland, which, in his eyes, had ceased to exist. Globocnik assembled his staff and drew up his plans for the expulsion from the area of 110000 Poles, after expropriating their property. Confident of Himmler's backing, he thought he could ignore the plans of another powerful Nazi boss, Hans Frank, the head of the General Government and Hitler's personal representative. Frank's ideas for the future of Poland were known to him, but he believed himself to be in a strong enough position to ignore the administrative hierarchy, though, like everyone else, he was well aware that Frank was not to be trifled with if anyone in his domain dared step on his toes.

Globocnik, like Himmler, was wrong. Frank had his own plans for developing these territories with the aid of German settlers; he had announced on 25 January 1940 that he was going to deport a million Poles, 'his' Poles, and now his hackles were up and he did everything possible to thwart the plans of the amateur archaeologist, even to the extent of making personal representations to the Führer. But neither Globocnik nor the SS suffered any ill-consequences as the result of this squabble between highly placed gangsters; the price was paid by the victims of one of the appalling tragedies in Polish history.

Odilo Globocnik's name did not sound too well in the ears of his friends the race experts, so this gaol-bird who had escaped the gallows devoted himself with extra zeal to the task entrusted to him by Himmler. The 'return to ancestral roots' in a vast area densely populated by Poles had the additional advantage of enabling him to please the Reichsführer by producing evidence of former Germanic greatness. Globocnik's archaeological finds were confirmed by the Ahnenerbe ('Ancestral Heritage Society'), which had been founded in 1933 but became active only at the same time as the Lebensborn organization, at the end of 1935. This 'scientific' society established by Himmler specialized in racial research and took an interest in everything connected with the usages and practices of the Indo-Germanic race from which the 'blond Aryans' sprang who invaded the whole of Western Europe in the fourth and fifth centuries

A.D., including Britain. (Hence the Nazi inferiority complex towards the English, whose blood in their opinion was practically unmixed, above all unmixed with 'inferior' blood.) It was under the aegis of this organization, which became one of the most criminal of the Nazi régime, that so-called scientific experiments were carried out in the concentration camps.

Globocnik spent several months drawing up his plans for the 'resettlement' of the Lublin population. In his secret report he said:

> The Polish families in Groups I and II are to be separated from the rest of the population and sent to Lodz, where they are to be Germanized or sorted out . . .
>
> Groups III and IV include children and Poles over the age of sixty who will be accommodated in villages established for them. All the sick, and Poles under the age of sixty who are unfit for work, will also be taken to these so-called 'pensioners' villages'. The age limit for children is to be set at fourteen (and not ten), as the younger children are unsuitable for work in Germany . . . Some of the Poles aged between fourteen and sixty in Group IV will be sent to the Auschwitz concentration camp.

This plan was not even submitted to Himmler. Globocnik and his Gestapo, SS and police aides, aided and abetted by the specialists of the RuSHA, opened their 'emigration centre' on their own initiative. Those selected for Group I had to look unmistakably Nordic, of course; they were sent to Germany, and would see Poland again only if they passed their Germanization test. Group II had to be even more meticulously examined with a view to picking out those with just a few drops of the right kind of precious blood. Group III, consisting exclusively of 'Polacks' with 'the physiognomy of Slavs', were also to be dispatched to Germany, to work in factories and on the land, wherever labour was short. The sick, the aged and children were to be accommodated in a camp near Warsaw, pending agreement on their fate (since Hans Frank refused to have anything to do with them). The suspect members of Group IV were to be sent straight to Maidanek or Auschwitz. There was one rule that applied to everybody. Immediately after their arrival at the transit camp at Zverinec, parents and children – 30000 of them altogether – were separated. Parents and children who made any attempt to see each other were threatened with severe punishment, and serious breaches of discipline involved death.

The expulsion of 110000 Poles from the Zamosc area began on a freezing November morning in 1942 and was completed in the following spring. It has never been possible to establish what happened to the great majority of the 30000 Zamosc children. Nearly 5000 who were considered suitable for Germanization were taken charge of by the Lebensborn organization and the NSV. Here is one form that was filled in for the police records:

No. 475 Kuncewicz	
Surname and forename	Kunze, Andreas
Date of birth	4.11.1933
Place of birth	Zamosc Nationality D.R.
District	Poland
Marital status	Single
Resident at Rouffach since	20.8.1943
Father's surname and forename	Kuncewitsch Ceslaw
Father's denomination	Catholic
Father lives at	Poland
Father died at	—
Mother's name and forename	Kunze, Irene
Mother lives at	—
Mother died at	Zamosc, 26.5.43

This card, which miraculously escaped destruction, is one of the most valuable in the Lebensborn files. It was found by the French army when the town of Rouffach in Alsace was liberated. Not only is it an additional demonstration of the involvement of the 'L office' in the Zamosc affair, but it also enabled us to track down one of the few eyewitnesses. It sums up in a nutshell the fate of the children kidnapped in the name of the pure-blooded German race. The Pole Andrzej Kuncewicz became the German Andreas Kunze only a few days after being parted from his family. The Germanization of his surname was the prelude to total Germanization. The boy who was born at Zamosc in 1933 had been kidnapped in the spring of 1943 and sent to various reception camps before arriving at a State boarding school at Rouffach in annexed Alsace where the Lebensborn organization sent him to be indoctrinated, for he was too old for adoption. At this establishment not only German boys were introduced to their role of future masters of the world, but also sons of Germans living in the United States, Canada or South America whose fathers dedicated them to Hitler in the belief that he was going to win the war.*

But Andreas was a foreigner, a Pole. Never mind; the group of ten boys to which he belonged, eight of whom were pure-blooded and two 'valuable', were processed in double-quick time. That was how Andreas lived, until the liberation of Alsace by the troops of General de Lattre de Tassigny in 1945. When the boy returned to Zamosc, no survivors of his family were left to meet him. It was therefore with good reason that spaces had been left on the form for entering the father's and the mother's place of death. *The father* or *the mother died at* ... simply means 'was killed at ...'. Andreas's mother died on 26 May 1943 in unknown circumstances. Did Globocnik's men bludgeon her to death when she tried to see her child? Was she sent to Auschwitz by the race experts responsible to Ulrich Greifelt, the big chief of the RuSHA, with the comment 'unsuitable for Germanization'? Or did she die of starvation or suffocation in one of those ghost trains crammed with 'undesirables' from Zamosc whom Frank refused to accept, with the result that Globocnik ordered it to be sent 'somewhere else', whereupon Greifelt diverted it and Eichmann took it over?

Andreas never discovered. His father's fate remains equally speculative. Did he become a slave of 'decent employers' in the Greater German Reich, or was he shot for some act of resistance, revolt or insubordination? (Between 1940 and 1943, 8000 persons were slaughtered at Zamosc in a 'rotunda' specially prepared for mass executions.) Did he die at Treblinka? Was he beaten to death for approaching the

*The index-cards concerning these foreign Germans still exist. These American, Canadian and other 'Nazis' do not seem to have been called to account after the war for their act of fidelity to the Führer

barbed wire that surrounded the huts where the RuSHA was selecting children? Or was he shot as an 'example' in the market place with other hostages? Or did he die like thousands of his countrymen of cold, hunger or suffocation in a hermetically sealed train shunted on to a siding at Zamosc, Zverinec, Lublin or Warsaw and 'forgotten' for days on end?

Being over the age of six, Andrzej Kuncewicz, who became Andreas Kunze, did not have the honour of being given German adoptive parents in place of those he had 'lost'. That age-limit had been fixed by Professor Hildegard Hetzer, one of the Reich's leading public health officials, and a lady who continued to practise in the medical profession after the end of the war. Her duty had been to submit kidnapped children to psychological examination with a view to determining to what extent and within what period of time they might be turned into good Germans. It was she who received them at the camp at Brockau when they arrived in transport provided by the SS and the Wehrmacht. The involvement of the Wehrmacht is readily explicable. Between 7 July and 25 August 1943, 4454 children between the ages of two and fourteen were taken from their parents in the Lublin area and deported to Germany. Bearing in mind that three times more children were taken by lorry to Brockau in less than six weeks – the non-valuable having been already eliminated – the SS could obviously not have managed without army aid.

The other children, after being sifted and declared racially acceptable, were provided with forged birth certificates through the agency of Dr Tesch and distributed to Nazi families. These families, with few exceptions, kept them after the war, allegedly 'in the interest of the child, as it was an orphan'.

Children of seven or more were not so easily disposed of. After thorough interrogation by Professor Hildegard Hetzer's aides, some refused to admit that their mother was a Polish whore who died of tuberculosis or that their father had been murdered by Polish bandits who were trying to rob him. The older children in particular were often well aware of the circumstances in which they had become orphans. Some had seen their parents killed, others had had the scene described to them by other children who had been forced to watch hangings in the public square.

Physically they were suitable for incorporation into the German nation, but their reluctance to fall in with the criminal behaviour of these Germans who claimed that they wanted to rescue them from their misery clearly pointed to their being mentally disturbed. Those like little Andreas who were nevertheless regarded as potentially Germanizable were sent to the *Heimschulen* (home schools), the boarding-school melting-pots intended to produce the future janissaries of the Greater German Reich. The incorrigibles who refused to deny their Polish origin or the version of their family history offered them by their interrogators left for concentration camps, or, at best, labour camps.

In November 1942 Globocnik's plans for the 'resettlement' of 110000 Poles were complete and his emigration centre at Zamosc became operative. Whole battalions of SS, aided and abetted by the professional 'disinfectors' of the Gestapo, the Sipo and the RuSHA, set about the task of expelling these people who had imprudently been living for centuries on allegedly German soil. Parallel with the plan for their expulsion was another to establish German settlers throughout the territory with the aid of a rational distribution of property confiscated from Poles and Jews. The intention was to send 85 per cent of the Polish population and 65 per cent of the population of the Ukraine to Siberia and within ten years to settle just under three million Germans in the expropriated territory, which would be administered by Himmler and the SS.

The winter of 1942–3 was one of the most severe of the war. Peasants thrown out of their homes were taken in trucks or on foot to be sorted into Groups I, II, III or IV. The sorting process, the separation of families, the seizure of children, rapidly created a climate of terror, one of the most horrible ever known in Poland, where at the end of the war there were 359 empty villages, all the inhabitants of which, including children, had been exterminated. But, for the Poles of Zamosc, cold, hunger, the loss of their homes and their land, the agony of losing those dearest to them, were soon supplemented by total administrative chaos. Squabbles and jealousies between the various Nazi agencies led to the greatest shambles in the history of deportation, for which the 'non-Germanizable' population of Zamosc paid dearly. Thousands of men, women and children were herded into squalid transit camps pending orders. Many died of cold and hunger in the Wehrmacht trucks taking them to slavery in Germany, the intended destination of Groups I and II. That of Groups III and IV was 'villages for retired persons, in other words extermination camps. Some thousands of children who had the luck to have blue eyes and fair hair were also packed off to Germany after a sifting or two. But no one knew what to do with the remainder, piled into railway coaches with the aged and the sick at the nearest station. These coaches were sealed and shunted into sidings pending orders from Greifelt, Globocnik or Eichmann, or counter-orders from Frank or the Wehrmacht High Command, or even from Globocnik himself. Meanwhile nobody cared, for all Globocnik was interested in was recovering territory, while Lebensborn and VoMi were preoccupied with recovering valuable children, and all that Frank and his thugs cared about was ensuring that no more expelled Poles arrived on his territory.

Nevertheless, every so often somebody somewhere would give an order, and it would be decided that at such-and-such a time and such-and-such a place a train would be leaving for Auschwitz. The race experts would point to the 'very valuable', the 'relatively valuable', and the rest, who would be sorted accordingly, and the old, the sick, the 'recalcitrant', and mothers with children in their arms would be herded into the overcrowded compartments. One day these gentlemen selected a party of 200 'Eastern-type' children with their mothers. They were told to prepare themselves for a medical examination. The whistle blew and off they went. There were no survivors.

In spite of orders, counter-orders and destinations that were changed at the last minute, more and more overcrowded trains accumulated at stations in the Lublin area which acted as collecting points, but they were forced to wait by the Government General's refusal to accept more displaced Poles over and above the million Jews and undesirables with whom it was already encumbered, or, if they did leave, they had to go back on their tracks. Meanwhile it grew colder and colder, Russian military pressure steadily increased, and the Wehrmacht, which was having to send reinforcements in the direction of Stalingrad, had priority on the roads and railways. But Globocnik was not to be diverted, and early one morning in the freezing wind that swept the Lublin plain a train left for Warsaw, on the way to Poznan and then Germany. The passengers were children. The news spread through the region, and at every station a crowd waited, hoping it would stop, for they wanted to save these children, or at least give them bread, milk, potatoes, blankets. It stopped once only, in the middle of the night. The crowd rushed it, forced the doors and managed to grab a few children from the groaning mass of humanity. Sometimes a child thus extricated was already dead and was put back while hands went on searching for a living body. One such body was that of an old man. 'No, leave me, take a child,' he said. The SS guards, who had been taken by surprise, opened fire, and the crowd fled, taking its booty, a few moribund children, none of whom survived.

The train started up again. At each station it slowed down as if it were going to stop in front of the crowd waiting to take it by assault, but then it accelerated again. Inside the coaches the cries intensified, while the helpless crowd raised its arms to heaven and looked at the passing faces of starving children yelling through the barred windows. How many of those who spent several weeks in those railway coach-cages survived? No one in Poland knows.

One December night the death line was sabotaged by partisans near the village of Piaskie-Ruskie. Next day

Globocnik's men shot a hundred mothers with children in their arms in the village square. The ghost trains promptly resumed moving towards their unknown destinations, and so it went on until the 'disinfection' of the region was complete.

To speed things up, the babies of the village of Lomza were taken into the forest and 'disinfected' by a hail of hand-grenades. Older children not sufficiently Aryan in appearance were sent to forced labour in the children's camp at Lodz. More than a thousand children aged between three and twelve were interned there as anti-Fascist war criminals.

Special cases were dealt with personally by Globocnik. These consisted mostly of individuals 'of good Germanic stock' who in spite of threats, blows and deprivations refused to raise their right arm to the swastika flag. The SS devoted a great deal of its precious time to about a dozen of these recalcitrant persons or persons of mixed blood who might 'constitute a danger to other Poles of Germanic stock in the Zamosc area'. Globocnik, whose intelligence was not above the average required for the SS, was the more relentless in dealing with these people because, among other things, they were intellectuals, a species especially loathsome to representatives of the new order.

A voluminous correspondence was exchanged between the various police agencies with regard to this handful of ethnic Germans, who were the subject of prolonged investigations. The results were submitted to Himmler, who on 10 February 1943 addressed this reply to the RuSHA, the VoMi and others concerned.

I enclose particulars of five Poles or Polish families of German stock.

1. Re Johanna Achidzanjanz of Tomaszow. She is of 50 per cent German stock, but is totally Polish-minded. She is to be transferred from Zamosc, which is in a German settlement area, to within the area of the General Government.

2. Maria Lambucki of Tomaszow-Lub is of 100 per cent German stock and is totally Polish-minded. She has completely

repudiated her Germanity. I direct that she be transferred immediately to the Ravensbrück concentration camp. Both her sons, who are aged eight and thirteen and are racially very good, are to be sent to Germany by the head of the RuSHA with the co-operation of the security police and are to be separated and sent to different especially well-managed home schools [Nazi boarding schools for ideological indoctrination]. I require quarterly reports about their progress. Correspondence with their mother is forbidden until further notice, until she realizes the treasonable nature of her conduct. The teachers and fellow-pupils of these two boys of good race must do their utmost to make it clear to them that they are in no way disloyal to Poland, but by reason of their antecedents and racial value owe loyalty exclusively to the blood from which they sprang. This must be the overriding endeavour in all such work, namely to win back valuable racial elements that have politically and nationally gone astray.

3. Stanislaus Koch, from the Sitno domain, is of 75 per cent German stock. He repudiates any association with Germanity. He is to be transferred to a concentration camp for work in an armaments factory; so are his wife and daughter. If practicable, they are to be put in different camps.

4. Brunhilde Muszynski, *née* von Wattmann, is to be sent to a concentration camp. Her father's origin is to be checked because of a suspicion of Jewish blood. This is to be done as quickly as possible. If the family tree turns out to be in order, the two children aged four and seven are to be placed by the head of the RuSHA in a German family, or alternatively the older child might be placed in a home school.

5. Ingeborg von Avenarius, *née* Wattmann, is to be sent to a concentration camp; her children are to be dealt with by the head of the RuSHA in the same way as the other children.

6. In all the above cases the head of the RuSHA will submit proposals to me, enclosing photographs, racial evaluations and a report on the impression made by the individual child and its mother. I will then make my decision in each individual case. Until further notice both parents and children are forbidden all correspondence.

One hundred and ten thousand persons were expelled from the Lublin district, their destination being either the gas chamber or forced labour in Germany; 30000 children died of cold and hunger, sometimes only a few hundred yards from what had been their home; and nearly 5000 children were taken in charge by the 'Brown Sisters' of the NSV and the representatives of the Lebensborn organization, while Globocnik persisted in his aim of clearing his area of sub-races as quickly as possible and Frank raged and threatened, wrote indignantly to the Führer, and did his best to thwart him. And while all this horror was going on, Himmler busied himself with the fate of a dozen individuals.

Kaltenbrunner, Heydrich's successor, also devoted a great deal of his time to this relatively minor matter of recalcitrant ethnic Germans. After carefully studying the matter, he sent a long report to Himmler on 24 April 1944. The Reichsführer, who might have been expected to have other preoccupations at that moment, read it with care on 17 May. He approved of the first, second and third pages, which were more or less a résumé of the ideas contained in his own letter of 10 February of the previous year. But page 4, containing Kaltenbrunner's conclusions based on those of the Lublin RuSHA, engaged his special attention. The questions which follow are Kaltenbrunner's; the italicized phrases are those Himmler underlined, with his comments in square brackets:

I request a decision on

1. Whether Maria Lambucki, though we have an interest in regaining her family for the German people and though she was granted a certificate of German nationality on 4 March 1943, *is still to be arrested.* [No]

2. In the event of the arrest of Maria Lambucki, what is to be done with her two sons? [This was crossed out.]

3. Whether *the arrest of Stanislaus Koch is still to take place*, though we have an interest in winning the family back for the German people and it was intended to resettle him on a farm in Germany. [No]

. . .

5. Whether Jachwiga Koch should be placed in a German home school in case Koch should not be arrested. [Yes]

Kaltenbrunner's letter continued:

As the Jewish descent of *Brunhilde Muszynski* has been established and her arrest approved, and as the sending of her children to German home schools is impossible in view of their Jewish descent, I further request a decision on what is to be done with the *four and seven-year-old children.* [To be sterilized]

Kaltenbrunner also wanted to know where the two children of Ingeborg von Avenarius were to be sent, since her Jewish origin had been established and her arrest ordered. Himmler's answer was 'anywhere'.

17. The Nordics of Lidice

'The non-valuable children will be put in special camps. The children of good race, who obviously could become the most dangerous avengers of their parents if they are not humanely and correctly brought up, should . . . be admitted to a Lebensborn children's home for a probationary period, where as much as possible about their character should be discovered, and then be sent to German families as foster-children or adopted children.'

HIMMLER to Max Sollman, 21 June 1943

To a careful administrator like Heinrich Himmler writing 'to be sterilized' at the bottom of a page was merely a matter of routine. On the other hand, when we met the aged and paralysed Dr Ebner it was difficult to imagine that thirty years earlier this impotent old man had personally ordered the sterilization of young children. This short, bald, obliging Bavarian country doctor had been an excellent obstetrician and was very popular with his 'nurses', among whom he had the reputation of being 'great fun'. After the war he returned to his practice in the village of Kirchseon in Bavaria, and the German Red Cross awarded him a medal for his services. The documents dealing with the kidnapping of Rumanian children for the benefit of the Lebensborns show, however, that the orders for the sterilization of 'non-valuable' children were signed, not by Himmler, but by SS Oberführer Gregor Ebner.

The surviving 'L office' documents about the abduction of children in that country do not enable us to state how the kidnappings took place or how many children fell into the hands of the SS. Apart, of course, from its Jews, Rumania, being an ally of Germany, was spared the worst racist excesses of the Nazis. An indeterminate number of kidnappers seem, however, to have worked in some parts of the country without the knowledge of the local authorities. Evidence is provided by a letter from the VoMi to the Lebensborn headquarters, Berlin, dated 14 July 1941, informing Dr Ebner that the number of children from the Banat region had increased from thirty to 185; that a number of those children were said to be under the age of six; and that the exact number of boys and girls had yet to be established.

On 25 August Ebner confirmed to Sollmann that twenty-five children from the Banat region had been sent through the VoMi to Langenzell Castle. It is anybody's guess what happened to the others. Ebner wrote:

> From the racial viewpoint, only a few of these children are to be regarded as an enrichment of the German people. Altogether only two children are stated to be suitable for adoption. The eighteen children unsuitable for adoption by reason of their age should be placed with foster-parents or handed over to the Labour Service. Five others are stated to be totally useless, for racial and biological reasons; one of these, Agnes Fiala, should be sterilized immediately, as the young fellows in the camp are said to be already beginning to take an interest in her. Also two of the boys should immediately be made incapable of reproduction, one of them, Nikolaus Reiszer, because he has tuberculosis, and the other, Georg Kuhn, because with his protruding ears and round shoulders he makes an impression of degeneracy.

A few days after Dr Ebner visited Langenzell Castle the two 'suitable' children were transferred to the Taunus home near Wiesbaden to await adoption. The others who were not to be sterilized were placed in German families or sent to labour camps, if they did not end up by being enlisted in the Waffen SS. The sterilization operations took place in a hospital at Heidelberg.

An SS man named Staudte stated in evidence at the Nuremberg trial that Himmler, while on a tour of inspection in Rumania, took away with him in his car a fair-haired, blue-eyed thirteen-year-old boy named Camillo Boehna, who was first placed in a Lebensborn home and then adopted by a German family. This youth has never been traced. Rumour has it that Himmler one day became 'ecstatic' at the sight of two blond Ukrainian boys whom he also had transferred to Germany. True or false, the story obstinately persists.

On 11 July 1941, immediately after the German assault on Russia began, Himmler published an order of the day in which he instructed the SS to lay its hands on children of German extraction. A few months later a report landed on the Reichsführer's desk of an experiment in 'the selection of thirty children' with a view to their Germanization, carried out secretly at Bobruisk in central Russia by the Pflaum special squad.

In May and June 1943, during the so-called Operation Winter Magic against the partisans in North Russia, the commander of the Sipo and SD in the East European countries reported that:

> 1000 children have been evacuated and some of them, after separation from their parents, have been accommodated in children's homes in Riga and Libau Those who could not be accommodated have been provisionally entrusted to Russian families. In so far as these children remain in Lithuania, they will have to be permanently left with these families, which, however, will mean that the Russian racial potential will be strengthened in that country, because these children are naturally receiving a purely Russian upbringing and educátion.

In the offices of the German Red Cross at Hamburg in March 1973, we met a young woman whose fate may perhaps have been typical of the thousand children assembled at Riga after being separated from their parents. Larissa arrived in Germany with 'a Russian family', probably of German descent, who 'worked for the Germans'. She knows neither her name nor her age nor anything about her early life, and still has a pronounced Russian accent, though she has lived in Bavaria for thirty years. As she told us:

'My life is one long martyrdom. With no family, no mother, treated as an alien everywhere, I still suffer from not being German. For a long time I was called a dirty Jewess, and a doctor in a hospital where I worked in 1963 actually told me that in Hitler's time people like me would simply have been thrown out of the window.'

She applied to the German Red Cross in the hope of finding her mother, and since then a poster with her photograph has appeared from time to time on the walls of German towns, accompanied by the words: 'Who knows me? Who can tell me where I come from?' No one knows how many Larissas there are in Germany who are asking the same question. They do not know whether they come from Riga, from the Banat, from Bobruisk, or simply from a Lebensborn home, born of a 'slave' mother of good race who in exchange for a confinement in first-class conditions had to leave her child to the Führer. The total lack of documents means that their questions are likely to remain unanswered.

The fate of Russian children abducted by the special squads or taken away by Wehrmacht units during their retreat has never been established by any Western inquirer. A few isolated cases were reported in West Germany to the investigating commissions set up by the Western Allies during the post-war occupation of the country, but were immediately passed on to their Russian counterparts. So far as the present inquiry is concerned, various approaches made in the hope of getting some sort of reply to the questions raised in the documents quoted above remained fruitless.

Frau Marga von Mann, a general's wife, had for a long time wanted to have a child, and in March 1943 the happy event seemed about to take place, for a letter arrived from the Lebensborn head office in Munich informing her that if she went to the Kohren-Salis home in Saxony her wish would be fulfilled. She was delighted, for less than a fortnight had elapsed since she put in her application. True, Max Sollmann, who had a great respect for the hierarchy, could not resist the pleasure of obliging a general's wife. It was also true that, because of the recent influx, a number of homes were full, which enabled the adoption service to cope with the demand.

Frau von Mann arrived at the Sonnenwiese home at Kohren-Salis towards the end of the month. She was received with the honours due to her husband's rank by Herr Ulrich, the head of the home, and, after the obligatory 'Heil Hitlers', she was handed over to Erika Wittmann, the head nurse. A young employee, Fräulein Edelmann, accompanied them. In one of the big centrally heated rooms with big bay windows opening on to the countryside she inspected seven babies one after the other, and finally stopped in front of a little blond boy who, Erika Wittmann assured her, was a pure German Aryan, named Johann Altzmann, born on 24 April 1941. The general's wife said she would take him. Her pleasure in the little boy came to a sudden end three years later with the arrival of a Yugoslav liaison officer acting for the international repatriation organization. Johann had been stolen by the VoMi in Yugoslavia and handed over to the Lebensborn organization.

Himmler's order of 25 June 1942 had directed his troops to take reprisals against Yugoslav partisans and other 'bandits' in the Lower Styria and Carinthia regions. The order stated:

> These operations must be mercilessly directed at all elements of the population that support the partisans. The menfolk must be shot without exception, even if whole clans are involved. The children must be taken away and placed in the German provinces . . . I require reports on the number of children and on their racial value.

On 14 September of the same year, Obersteiner, Himmler's personal representative in Yugoslavia, reported to the VoMi as follows:

> In accordance with the Reichsführer's order of 25 June 1942 (Circular no. 323/42 (top secret)), the children of the partisans and rebels classified in accordance with the specifications of the RuSHA have been sent to VoMi camps. Children in Categories I and II aged from six months to twelve years will be handed over to the Lebensborn, Herzog-Max-Strasse, Munich, which will see to the further care or adoption of these children of good race.

At a conference held at Veldes on 21 September, at which

Inge Viermetz was present, a decision to this effect was made by the manpower division of the SS. A 'note' on the discussion says:

...after sifting has been carried out the children will be separated from their mothers, and it was pointed out that a definite instruction from the Reichsführer existed that children of shot and executed bandits must beseparated from their mothers, so that no irresponsible hatred will develop among these children. After racial sifting, children with a good capacity for Germanization will be handed over to the Lebensborn Society, which will arrange for the adoption of these children by pure German families. The rest will remain with VoMi... The mothers are expected to go to concentration camps.

This document, signed by Dr Kürbisch, the head of Division 1 (manpower), and shown to us by the Austrian resistance movement, is the only direct evidence among the papers saved from destruction of the fate of the mothers of kidnapped children. Previously we had only seen veiled allusions or euphemisms referring to the death or abduction of these mothers. True, the mother of the Kuncewicz boy from Zamosc 'died' on a date corresponding to the massacres that took place in the area, but no information about her death is to be found. Other mothers 'moved to addresses unknown', which merely means that they were expelled, but nowhere else, either in Poland or in Germany, did we come across the phrase *Die Mutter kommen vorraussichtlich in KZ-Lager* ('The mothers are expected to go to concentration camps').

Inge Viermetz, as official representative of the charitable organization of Lebensborn at the conference at Veldes, subsequently paid several visits of inspection to the VoMi camps, looking for children 'suitable for Germanization', with definite instructions from Sollmann to 'take only young children who have not yet reached school age'. And, according to a Lebensborn employee at Nuremberg: 'A real competition for these children arose between the VoMi and the Lebensborn, and it was thanks to Frau Viermetz's efforts that the Lebensborn won in the end.'

It is true that Himmler's order of 25 June 1942 caused

embarrassment to the Lebensborn authorities because he mentioned children aged from six months to twelve years, while hitherto, until the beginning of 1942, only children under six could be accepted by the homes with a view to adoption. This embarrassment is confirmed by a letter from Dr Tesch to Brandt, asking whether – *à propos* of children kidnapped in Bohemia and Moravia – the 'older children of whom the Reichsführer speaks are really to be taken over by the Lebensborn'. But the question was quickly settled, as it had been or was to be in Poland, Russia, Rumania and Czechoslovakia. At an interview with Karl Hermann Frank, the Reich Minister responsible for Bohemia and Moravia, Sollmann agreed 'in certain cases to accept children up to the age of twelve, even at the risk that difficulties might ensue'.

While Polish abducted children were classified under the letter O (for *Ost*, 'east') the Yugoslav children were classified under the letters SO (south-east'). They were chosen entirely on the basis of their physical appearance, and had no papers. Dr Tesch provided them with new German identities, aided by Dr Ebner, who decided roughly when they were born. Often he was several years out, which led to a lively correspondence between the 'L office' and the adoptive parents. Some of this correspondence survived, and some of the families concerned gave evidence at the Nuremberg trial.

One of the instances quoted at Nuremberg was that of Frau Ilse Brauer, to whom little Hedwig Paczkowska was sent by the Lebensborn. She noticed that 'the child was much younger than was stated in her papers', and informed Dr Tesch, who referred the matter to Dr Ebner on 24 July 1943. Without admitting that he might have made a mistake on first examining the child, he wrote to Frau Brauer on 26 July and 11 September 1943 asking to be informed of the child's height and weight. It seems safe to conclude that during the years of unbridled abduction Ebner was sometimes so busy that he did not have time to examine in person all the children sent him for adoption. Once the children had been given new Germanic 'but not Nordic' names, and had been placed in good German homes, the Lebensborn tried to avoid all correspondence about them, so that all trace of their origin should disappear as quickly as possible. Their efforts in this respect were largely successful for, with very few exceptions indeed, both the Yugoslav investigators working for UNRRA in the immediate post-war period and we ourselves thirty years later found that the trail of these abducted children had gone cold.

SS Gruppenführer Karl Hermann Frank, an ethnic German from the Sudetenland, who took up his appointment as Minister of State for the 'Protectorate' of Bohemia and Moravia three days after its creation, was unrelated to his namesake the Governor-General of Poland, but was just as brutal. He was the butcher of Lidice.

Maria Hanfova, who was renamed Marga Richter, was no. 60 on the list of children under sixteen drawn up by the Gestapo at Kladno on 11 June 1942. The list includes the names of ninety-one children, only nine of whom have ever been traced. Maria Hanfova, who lives in the new village of Lidice, described to us the 'medical examination' to which she was subjected by Frau Professor Dr Hetzer in a RuSHA office at Lodz:

'We left Lidice on June 10 and arrived on the 14th at the Kinder-Ausleselager [camp for selected children] which was run by a woman [Frau Hetzer] who examined me herself. We were made to undress completely, girls and boys alike. The lady measured our heads, chests and hips. Then we were weighed one after the other on some big scales in the corner. After that we were photographed, the face only, but from all angles. We were half dead with fear, and above all without news of our parents, our village. We had been shut up in the school at Kladno, with our mothers. Then the SS separated us, and we were taken to the station. We knew nothing about what had happened at home.'

What had happened was reported by Horst Böhme, the

security chief at Prague, in the telegram he sent on 12 June to his SS colleagues at Lodz:

By order of the supreme command the community of Lidice in the Protectorate of Bohemia and Moravia has been rased to the ground in connection with the attempt on the life of Gruppenführer Heydrich. The whole male population has been shot. The women have been sent to a concentration camp for life. The children were examined for their capacity for Germanization. The non-Germanizable will be sent to you, and you will send them on to local Polish camps. There are ninety children, unaccompanied by members of their families. They will be sent to Lodz in a special coach attached to the regular passenger train to Lodz. The train will arrive on Saturday 13 June 1942 at 2130 hours. I therefore request you to ensure that the children are met at the station and immediately sent to a suitable camp. The children are in the following age groups: five between the ages of one and two; six between two and four; fifteen between four and six; sixteen between six and eight; twelve between eight and ten; thirty-six between ten and sixteen. They have nothing with them except what they have on their backs. Special care is not required.

This last phrase was equivalent to saying: 'They can drop dead.'

Heydrich, who had set about pacifying Czechoslovakia by terror, had had an attempt made on his life by two members of the Czech resistance who had come from London for the purpose. That had happened on 27 May while he was on his way to work at the usual time in an open Mercedes. His death several days later was the signal for bloodthirsty reprisals, which reached their climax with the decision to wipe Lidice from the map announced in Berlin by Karl Hermann Frank after a conversation with Hitler on 9 June. That same evening the village was encircled by detachments of police, Gestapo, the Wehrmacht and Military Police. The operation was directed by Böhme, the security chief, aided by qualified Gestapo agents such as Geshke, Skalak, Faber, Petrat, Burger, Felkl and their chief Wiesmann, most of whom were executed after the war. In a single night of terror all the males over the age of sixteen were shot, the houses pillaged, and the village burnt down. After the war

Weismann said: 'The men of Lidice went to the wall slowly but with head erect and with courage. No judgement had been read to them. They were shot without being told the reason for their execution.' One hundred and ninety-two men and seven woman were shot, including nineteen who were not at Lidice that day but who suffered the same fate in Prague a few days later. The other women were separated from their children and sent off to Ravensbrück. Next day Radio Berlin announced that Lidice had ceased to exist.

Maria Dolezalova, no. 44 on the Gestapo list, said:

'We spent three days with our mothers in the gym of the secondary school at Kladno. On the third day the Gestapo leaders arrived. The chief of them immediately insisted that the women, who had a long and tiring journey to a labour camp ahead of them, should let the children go first by bus, so as to be able to stand up to the journey better. He also insisted that the mothers should hand over their children voluntarily. Obviously none of them were willing to be parted from their children, so he threatened to shoot all those who refused. To show he meant business, he fired a shot in the air with his revolver. This of course caused panic among the mothers and children. The Germans took advantage of this to go for the mothers and snatch us from their arms.

'Believe me, that was a moment that none of us will ever forget, even in forty or fifty years' time. It's like a horrible, brutal film that keeps on passing before our eyes.

'After that horrible separation we were taken to a classroom, where particulars were carefully taken of the colour of our eyes and hair and our height and weight, and finally a card was hung round our neck. At the end of the third day they took us by truck to Prague station, and from there by train to Poland, to a town called Lodz. There we were put in a disused factory that was being used as a transit camp, and racial examinations took place. The commandant of the camp was Hermann Krumey, of the SS. There were, I think, eighty-nine of us children from Lidice.

'On the day of our arrival in the factory the Germans ordered us to sit down. Then they walked amongst us, pointing us out. Of course we did not know why. They chose six girls, including me. Afterwards we learnt that the selection was made by the representative of the Lebensborn organization at Munich. Other children were later added to our group. This was the group destined for Germanization . . . In July 1943 a childless German family from Poznan came to fetch me. This was the family of Alfred Schindler.'

Altogether thirteen children from Lidice sent from Prague to Lodz in June 1942 were considered suitable for Germanization. On 2 July they were sent to the Lebensborn home at Pushkau, where they were taught some scraps of German and the basic principles of Nazism. These 'orphans of German stock' were then handed over for adoption through the Lebensborn organization. Himmler himself charged Sollmann with taking the necessary steps. On 21 June 1943 he wrote a letter to Sollmann under the heading of 'secret', in which his attitude to these children of 'bandits' is clear:

> Dear Sollmann, I want you to get in touch immediately with SS Obergruppenführer Frank in Prague. The best thing would be for you to go and see him. The question to be solved is that of the care, education and accommodation of Czech children whose fathers or parents have had to be executed as members of the resistance movement. The decision must be a very wise one, of course. The non-valuable children will be put in special camps. The children of good race, who obviously could become the most dangerous avengers of their parents if they are not humanely and correctly brought up, should, as I see it, be admitted to a Lebensborn children's home for a probationary period where as much as possible about their character should be discovered, and then be sent to German families as foster-children or adopted children.

The placing of the Lidice children was the occasion of a voluminous correspondence between various SS agencies and the Lebensborn, particularly between the head of the home and Inge Viermetz, who seems to have taken a special interest in the matter. Sollmann, Tesch and she went several times to Pushkau to examine the children and decide to what kind of Nazi families the Lebensborn might entrust them.

The race experts, in their anxiety to catch every drop of

good blood that might possibly flow in the veins of a 'bandit's' child, carefully followed the progress of the six pregnant women from Lidice. Orders were given that they should be well treated until after the confinement. The Gestapo allowed their children to remain with them for three weeks, long enough for the race experts to decide on the child's future; then it was taken away, and the mother remained alone in the Pankrac prison in Prague. Later she was to meet her comrades again in the concentration camp of Ravensbrück.

Thus six children were born in a Gestapo gaol. They were given German names and handed over for adoption. Five died as the result of the bad treatment inflicted on their mothers. The survivor was a little girl who was traced after the war.

One pregnant woman from Lidice managed to pass through the Gestapo net. Her child, like Lucyna Gruszka of Rogozno, was born in a hut at Ravensbrück. But the camp guards discovered the child; it was taken away and never heard of again.

The final balance of the Lidice massacre was 192 men and seven women shot, 196 women sent to Ravensbrück, where forty-three died as a result of torture and ill-treatment, 105 children abducted, of whom thirteen survived. The Polish Commission for the Investigation of Hitlerite Crimes was never able to establish the exact number, for many children became 'orphans' when they were taken from their parents, who were executed or died after being deported. The eighty-two children who were regarded as unsuitable for Germanization were sent from camp to camp and ended by being forgotten and dying of cold and hunger. Krumey, who was head of the Gestapo at Lodz, wrote about them as follows to his chief, Eichmann:

> Since I have no information either from IVB4 or from the commander of the Sipo and the SD about the further disposition of these children, who were sent here without luggage, I urgently request instructions about their further disposition.

In the Nazi vocabulary of the time the term 'further

disposition' meant the death sentence, particularly if it were used in a telegram from Krumey to Eichmann about 'valueless' children. However, Eichmann still had the last word. Further telegrams followed, but none contained a clear order settling the fate of these luckless children. Eichmann when questioned during his long trial in Jerusalem said he had no precise memory of the case. After 30 July 1942 the trail of these 'bad' Lidice children disappears. All that remains is a few photographs and a letter, written in June 1942 at Lodz in Poland by Maria Sroubkova, no. 86 on the list drawn up by the Gestapo at Kladno. She was fifteen years old.

Dear uncle and aunt, we have been here for a fortnight already and we have not yet had a change of linen. Please send us linen, needles and thread, and a few German coins. And send us some bread, even if it's only the crusts you give the rabbits at home. Also tell us what has happened to papa and mama.

PART THREE

Thirty Years Later

'A great task awaits us after the conclusion of peace. We shall colonize. We shall hammer the laws of the SS into our young people. It is up to them to ensure that the élite of the German people will be fruitful. In twenty or thirty years' time we shall be the ruling class of Europe; we must succeed in giving the German people the necessary living space and in controlling and ruling all the countries of Europe. The end of the German people would be the end of beauty, culture and creative force on earth. Let us think of our Führer Adolf Hitler, who has created the Greater German Reich and who will lead us into the future of the German people.'

Himmler, addressing SS commanders on the Eastern Front, winter 1943.

18. The Collapse

'Praise be to God! The diabolical work of the Lebensborn has at last been destroyed.'
CARDINAL FAULHABER, speaking in the Marienplatz in Munich, spring 1945

After the Allied landing in Normandy in June 1944, the children born in the Lebensborn homes were evacuated to Bavaria in an atmosphere of total panic. On 23 November 1944 Ebner reported that all the homes in Holland, Belgium, Luxembourg and France had been abandoned. In the East the advance of the Red Army put an end to Himmler's dream of populating Europe with men and women of pure Aryan blood and high racial value. Doctors, nurses, pregnant women, babes in arms or just able to walk, were hurriedly loaded on to SS trucks or special trains and sent to Steinhöring in Bavaria.

Pierre Variot, a member of the French resistance who was deported to Dachau and worked in the 'Lebensborn squad' at Steinhöring, still clearly remembers the babies' arrival at the village station.

'It was in the autumn of 1944 that the first babies arrived. The first day was dreadful. We were used to being taken to the station by the SS to unload all sorts of things to be delivered to the home: sugar, coffee, cocoa, oranges, fresh fruit, clothing, etc. But that morning a Polish comrade handed me something wrapped in a blanket, saying "*Kind, kind.*" I don't know German, so I didn't realize that *Kind* meant "child". I took the package, and I was just going to throw it to the man by the truck when I heard a whimper. I was so startled that I nearly dropped it. I was also afraid because to my mind and according to my own experience, the SS could do nothing but kill, kill . . . My hands started trembling, and my legs too. I walked a few paces and handed the child to my comrade. The unloading took several hours.

Every so often the SS drove away with the truck, which came back empty. We wondered what they were going to do with the babies. We never knew, but at the time it was terrible, because we could not imagine that the SS could do anything but kill them.'

During the winter of 1944–5 Steinhöring, the model home directed in person by Dr Ebner, became the last refuge of the children of the super-race born at a time of military disaster. The reason why Steinhöring was chosen to shelter the 'biological substance' of the German people when there were many other maternity homes in central Germany was supplied by Dr Kleinle, a medical officer in the Wehrmacht, whom the Americans put in charge of the home when they took over.

'The house was in the heart of Bavaria which, so the SS believed, was to constitute the famous Alpine redoubt in which Nazi Germany would be able to hold out to the end of time. They proclaimed that the Alpine redoubt would never surrender, and that was why the Steinhöring home had to survive.'

Paula Hessler, the nurse who still lives in Steinhöring, also took part in the retreat of the Lebensborn women and children when they were evacuated from the home at Bad Polzin (where she was employed) in Pomerania.

'As the enemy advance continued, we had to leave the home in a hurry. The women and children, accompanied by doctors and nurses, travelled right across Germany by special train until they reached Steinhöring four weeks later. It was a great strain for the small children and pregnant women. Two women gave birth during the journey. Fortunately we lacked nothing, on the contrary, we had more than enough to eat, so that our train was several times stormed by hungry German refugees. The SS escort had to repel them by force.'

To escape the Allied air raids on Munich the whole staff

of the Lebensborn headquarters also took refuge at Steinhöring. The lawyer Dr Tesch set up his headquarters in the home of Father Koeppel, the village priest, and Sollmann lived in a flat in the big villa that housed the maternity home. The remainder were lodged in huts specially built in the park by inmates of the Dachau camp. When the offices in the Herzog-Max-Strasse in Munich were destroyed by bombs, the files were taken to Steinhöring. They were stored everywhere, in the huts, in the maternity home corridors, in Tesch's and Sollmann's flats. These two still believed that the Alpine redoubt would hold out, and that Germany would still win thanks to the secret weapons which the Führer had promised.

By the spring of 1945, the Steinhöring model home, which during the heyday of planned propagation had been intended for the accommodation of about fifty mothers, was in a state of chaos. More than 300 babies and young children, women on the point of confinement, dozens of nurses who had fled there from all over Europe, to say nothing of a large number of SS officers and men along with members of the Lebensborn administration, shared the villa and the huts. Military defeat added to the general nervous irritability, and life there became more impossible with every day that passed. The arrogance of the SS officers, who had not yet given up their uniforms, did nothing to calm the hysteria of the mothers or expectant mothers, whose numbers increased daily. Girls of 'pure German blood' resented the Norwegian, French, Dutch and Belgian women who had arrived there a few months earlier. The 'nucleus of the Germany of the future' was reduced to the state of a dirty and overcrowded camp. There were violent quarrels, and fights broke out over a baby's cot or an armchair by the window. Witnesses recall that indescribable chaos prevailed with women in labour and screaming children on their pots everywhere.

If these witnesses are to be believed, this chaos seems to have left the heads of the organization cold. With the exception of Dr Ebner, who had to deal with confinements almost daily, as well as with a phenomenal amount of illness among both women and children, their chief preoccupation seems to have been saving their own skin, though this did not stop them from helping themselves to the property of the organization. As Father Koeppel told us:

'For days on end the SS people drove to and fro between Steinhöring and some unknown place, presumably Hitler's 'eagle's nest' at Berchtesgaden. They drove off with truck-loads of food, linen, cutlery, everything of any value. When they finally left, there was not even a sack of potatoes left to feed the women and children.'

Baron von Feury, a member of the Bavarian Landtag and president of the Bavarian farmers' union, confirmed what Father Koeppel had said.

'I still remember very well the day when the Americans arrived outside the Lebensborn. They were going to take the place by storm, but the SS people had long since fled, taking with them everything that was of any value.'

Michael Brandmeir, who had been the Lebensborn store-keeper in Munich, said at the Nuremberg trial:

I saw with my own eyes how Sollmann, Tesch, Frau Viermetz and others picked out the most beautiful furs, the best bed and table linen, as well as pictures, carpets and food that had come from France or Poland. I actually had to take these things to their flats. The 'small' employees got their share too, although they had to pay their chiefs for it. But the important gentlemen kept the best things for themselves.

Today a Persian carpet covers the floor of Sollmann's living-room. 'You see that carpet,' he said to us. 'The Americans took it away from me, like all the rest of my property. Look at the state it's in. It looks as if a whole battalion had marched over it. Dreadful! And my books. They burned at least a thousand of them. And I had such a fine collection.'

Dr Kleinle, who made an inventory for the Americans, said:

'The linen, cutlery and furniture that the SS left in the house came from all over Europe. There was some cutlery with the Rothschild monogram, as well as table and bed linen from big hotels owned by Jews, as well as other things that had been Habsburg property, and so on.'

With all these preoccupations they had not been too busy to see about arranging for the disappearance of all traces of the identities of SS orphans while they prepared for their own flight. Father Koeppel remembered those events very well.

'For days the SS leaders had big bonfires lit in the courtyard and round the huts in the park. That was just before the arrival of the Americans. Presumably they destroyed in the process the birth certificates and other compromising documents that might have given information about the children's parentage.'

Paula Hessler, the only nurse whom the Americans permitted to remain at her post, because she was a real nurse, confirmed the old priest's memories. 'They burnt everything,' she said, 'the SS wanted to destroy the documents at all costs, otherwise why should they have burnt so many papers?'

Father Koeppel described the scene he found when he went into the house a few days after the arrival of the Americans.

'When I entered the house a few days later the floor was strewn with papers of all sorts; there were scraps all over the place, in the huts, in the yard, and even in the neighbouring fields. In their hatred of these Lebensborn people, the villagers wrecked the place and smashed to bits everything they could lay their hands on. If you realize also that American tanks drove along the paths and into the courtyard, you can imagine how the documents vanished.'

According to witnesses, the Lebensborn heads had suddenly fled during the last days of April. That morning hysteria must have reached a climax. Pregnant women and nurses, noting that those in charge had disappeared, rushed to the administrative offices on the ground floor, and found them empty. It was then that Dr Ebner appeared, in full uniform. He was the only one who had had the courage to remain at his post. A few hours later the first American tanks appeared at the gates and took up positions round the house. Paula Hessler takes up the story:

'The Americans knew there were SS in the village. They looked for them everywhere, but the only one they found was Dr Ebner. No sooner had they taken him prisoner than soldiers stood him against a wall to shoot him. I still remember the scene very well. A soldier, a black, had taken aim and was just going to pull the trigger when an American army doctor came into the yard and called out an order, whereupon the soldier lowered his weapon. Ebner, who was standing there with raised arms, did not immediately realize that the American doctor's intervention had saved his life.'

Sollmann, Tesch and the others were picked up several weeks later. They had simply gone back to live with their wives in the hope of being forgotten.

By order of the Americans, who occupied that zone of Germany, the last bastion of the Lebensborns was immediately converted into a children's hospital, and Dr Kleinle was put in charge.

'I arrived at Steinhöring on 15 May 1945. The place was in a state of complete disorder. There were German children there, as well as others born in the most various occupied countries. They were of all ages up to four. Altogether there must have been just over 300.* The few girls who had stayed with their children or were about to be confined were German, Norwegian, Danish, Dutch, Belgian and French. The foreign girls had a panic fear of having to go back to

*The Steinhöring home is mentioned here merely as an example. The SS had not been able to evacuate all the homes before they fled. At the Hohehorst home in north Germany, for instance, British troops were astonished to find fifty-one parentless children of two to five.

their country with a child born in a Lebensborn. We were able to persuade some of them to go back, particularly the French girls, who accepted the situation very bravely. But, though I remained at Steinhöring for several years, I never saw a single father. Not one man turned up at the clinic, not even when the girls were still there with their children. Most of the children, who were now completely abandoned, had to be distributed all over the world, the United States, South Africa, Britain, France, Switzerland, Denmark and also Germany, of course. The arrangements were made by charitable organizations. The first few months after the end of the war were very difficult, as we had no food for the children to eat and were short of staff.'

The 'Little Blonde Sisters' dismissed by the Americans were replaced by Salesian nuns from a neighbouring convent situated on a peak near the Austrian frontier. The SS 'nurses' had not made a favourable impression on the Mother Superior.

'Before they were thrown out by the Americans they treated us as if the SS were still the masters of the country. They were extremely arrogant. But not many of these "sisters", who were also mothers of some of the children in the home, remained; at most fifteen for nearly 300 children. We didn't know where the other mothers were, and we never heard anyone mention them. Nor did any father ever turn up to take his child away or to pay it a visit. That never happened.

'These children did not know what tenderness was. They were used to being in bed or living in groups, and were frightened of any grown-up who approached them. The few mothers who were left cared only for their own child and never about any others. The older children, the three- and four-year-olds, could not even talk. They merely expressed themselves onomatopoeically, like young animals. That is typical of children brought up in institutions. Also they were very backward in mental development in comparison with other children of the same age. One really felt very sorry for them indeed. Some were intelligent, others a little stupid and some were even mentally defective.'

When we asked Dr Kleinle whether he had detected any signs of a super-race, he replied:

'Most certainly not. The number of backward children among them was far above the average, about 15 per cent. They were feeble-minded or suffered from other disturbances. Those children were born in the last few months of the war . . . the disturbances only became apparent some months later.'

Most of these handicapped children were taken into special homes.

Thus Himmler's attempt to lay the physical foundation of a Nordic super-race with the aid of the Lebensborns, which functioned for nearly ten years, ended in abject failure. This applies both to the offspring of selected males and females as well as to the fair, blue-eyed children who were kidnapped all over Europe. So far as the former, the 'SS children', were concerned, to judge by the 'samples' we were able to examine thirty years later, they differed in no way, either in physique or in colour of eyes or hair, from people of the same age born in any other maternity hospital. As for those who were kidnapped, as they have grown older their features, shapes and hair colour have changed, like anyone else, and few of them today would stand much chance of being accepted by an RuSHA selection committee.

In 1946 the Lebensborn children had become 'orphans of shame', left to their fate and entrusted to anyone willing to take care of them. In a way they suffered the same sort of fate as their parents; they were shunned by society. They were distributed to organizations that hastened to wipe out all trace of their origin. They were creatures without a past. No one must know that their mothers were unmarried girls indoctrinated in Nazism and that their fathers were members of the SS.

For this reason, shortly after the war, town clerks were told to be 'understanding' about the details on the birth certificates; no mention was to be made of the SS, or of the father's rank in that fraternity.

19. Orphans of Hate

'We are the product of natural selection, the progeny of a race that is thousands of years old. Alien peoples have insinuated themselves into our midst and left us their heritage but, because of our blood, our people will triumph, for it is united by the sacred ties of Nordic blood. On the day we forget the foundations of our race, on the day we forget the sacred principles of selection and toughness, the germ of death will enter into us. We must always remember our doctrine: blood, selection rigour . . .'

HEINRICH HIMMLER, Poznan, 4 October 1943

An article in the Warsaw *Gazeta Ludowa* of 16 October 1947 stated:

Suspicious action by the Germans; getting rid of handicapped children.

The repatriation of children taken by force to Germany during the war is running into great difficulties. As has been confirmed, their number amounts to 250000. But so far it has been possible to bring only 2000 young children back to their homeland. A considerable percentage of physically and mentally handicapped children has been noted among recent arrivals. They were accepted without hesitation, on the basis of documents compiled in Germany since the end of the war. But thorough examination has shown that most of these children speak German only and were born illegitimately in Germany, often of unknown parents. There are grave doubts about the Polish nationality of these mentally handicapped children, and it seems as if the Germans may have used the opportunity to get rid of unwanted ballast. . .

On 4 November 1947 John Widdicombe, the delegate of the International Refugee Organization in Poland, sent a copy of this article to his headquarters at Heidelberg in the American Zone and asked for an inquiry to be made into these children's nationality. At that time, and even later, none of the IRO's investigators had any knowledge of the documents dealing with the Lebensborn organization, part of which had by then been recovered. So nobody realized that the Lebensborn, which took charge of kidnapped infants, would never have taken foreign handicapped children into its care, for did it not rid itself of its own Nordic handicapped children by sending them to Görden to be dealt with by the 'disinfectors'?

However, in the post-war files dealing with the search for kidnapped children, no trace is to be found of these handicapped infants sent to Poland by the Germans 'in order to get rid of them'. Thus the Eastern countries occupied by the 'Bolsheviks' stepped into the shoes of Görden. The occupying armies actually transported to those countries handicapped children who 'could not possibly be German'. The idea of the super-race and sub-humans had made progress and survived its originators; and there was no one among the occupiers who asked how it had come about that so many abnormal children assembled from all over Germany could be Polish and not German.

This was in 1947. The friends and allies of yesterday who had fought side by side against the Nazis, were slowly sliding into a war which was to last twenty years, the Cold War. Because of the repeated failure of the Allied occupation authorities to take away the adopted children of Nazi families, on the grounds that it was 'in the interest of the child' to leave it where it was, there was mounting tension between the Polish authorities and those of the Western countries. On 11 June 1948 the Polish newspaper *Zycie Warszawy* published an article under the headline: '200000 Polish children cannot return to Poland because of the criminal attitude of the Germans and the British occupation authorities'. It said:

The invasion and occupation of Poland by Germany cost six million lives. But Poland's biological losses are not limited to that figure, not only because it is impossible to assess these losses statistically, but also because the Germans stole from the Polish nation more than 200000 children in order to Germanize them.

This figure of 200000 includes the children stolen in Poland by the Race and Settlement Head Office [RuSHA] as well as the children born on Reich territory of parents deported for forced labour. This figure of 200000 represents only a rough estimate. It is in fact impossible to assemble the documents concerning the murdered parents, which were destroyed by the Germans before they left Poland. Fortunately some documents dealing with stolen children, crimes that the Nuremberg court described as 'crimes against humanity', have been recovered. They are in the possession of the Polish authorities, who for three years have been concentrating all their efforts on having these children returned from Germany.

Results of the efforts of repatriation: So far the Polish authorities have succeeded in securing the repatriation only of 20000 children from the Russian zone and – a fact which seems incredible – only 6000 from the British, American and French Zones of occupation. This figure constitutes only a small percentage of the children now living in West Germany.

The satanic reason: Why have so few children been repatriated from the west of Germany? It is nevertheless a historical fact that the occupier sent the largest number of deportees westwards, to the interior of the Reich. Thus there are great difficulties, but what do these difficulties really consist of? In the first place they lie in tracing the children and in the dishonesty of the Germans who do not declare to the occupying authorities that they are bringing up non-German children. There are also difficulties at the level of the children themselves who, demoralized and passionately Germanized, deny their nationality . . .

The diabolical reason lies elsewhere; it must be sought in the attitude of the British and American occupation authorities, and the British in particular. These authorities are not satisfied when a child is tracked down, when evidence of its identity is produced and even its parents (if they are alive) claim it. All that is not enough for them. They do their best to ensure that the child is not returned to Poland . . .

The criminal attitude of the British: The British nation, which has an ancient tradition of pioneering and has organized imperialist conquests for centuries, has an urgent need of pioneers, the new pioneers of British imperialism, and it is trying to find them wherever it can. At the conference of the International Refugee Organization at Geneva on 10 May 1948 it was decided that searches for children should cease in July 1949. Children found after that date will be sent to the British colonies. Thus there is no hope that in 1948–9 the British authorities will remove the obstacles hitherto encountered by the Polish investigators.

In the light of the Geneva decisions and in view of the attitude of the British authorities in Germany, it appears that the British not only tolerate German crimes against humanity – crimes that consisted of robbing a nation of its children – but also that it now co-operates with them. Moreover, they wish actually to profit from them.

After reading this article, the chief of the tracing services in the British Zone wrote to his colleagues in the American Zone that these reports 'are extremely exaggerated and can only be considered as pure Communist propaganda'.

It was in this atmosphere of mutual recrimination and suspicion, and of political propaganda, that the last searches for kidnapped children took place. Matters were made even worse by the fact that the defeated Germans did everything to hinder the searches. The hatred of the Communists inculcated by the race theorists in the form of hatred of Eastern sub-humans was intensified by the fact that it was in the East that the German army had suffered its most humiliating defeats. It was also because of the East that there was no longer 'one people, one nation, one Führer', but two peoples, two nations and no more Führer, since one part of divided Germany was 'occupied' by the Western powers, while the other had been 'stolen' by the Russians.

As at Zamosc, Lidice and Poznan, it was the victims who again paid the price of the conflict. Poland was in the East and, like Czechoslovakia, had gone over into the Communist side. The nations were nursing their own wounds and mourning their own dead. The vast numbers of refugees in the displaced persons' camps weighed more heavily at the time than a few thousand stolen children. On top of this there was the 'interest of the child' to be considered, that famous interest in the name of which they had been taken from their families in the first place. The British, American and French investigators, often motivated by the most generous feelings, hesitated to create new dramas in the minds of young children who could remember only their adoptive parents. Moreover, Poland was a poor, devastated, under-developed, Communist country. No one knew whether children who were being brought up in comfortably-off families would find similar conditions if they were sent to the East.

Thus the post-war files contain a multitude of reasons why *Ostkinder*, a great many *Ostkinder*, stayed in Western Germany. The Russians for their part made no exceptions for the children in their zone. With the efficient aid of the Soviet police, few East Germans defied the orders to return all foreign children living under their roof. The same orders were given by the Western Allies in their zones, but many West Germans took advantage of the division of their country into three zones to cover their tracks. The Polish Red Cross has hundreds of documents about families who 'left without leaving an address' just at the time when the International Refugee Organization began its inquiries. There were also a large number of cases of 'Germanized' children being left with adoptive families whose 'father' was still in prison for war crimes at the time the decision was made.

At Katowice we met Dr Roman S. Hrarbar, a lawyer, who in 1945–7 was head of the Polish mission responsible for repatriating Polish children. He said to us:

'In view of the lack of enthusiasm shown by the occupation authorities in helping us to find the children, we had to resort to very unorthodox methods and behave more like detectives or spies; and when we had tracked the children down other difficulties arose. How were we to prove the identity of a child who remembered nothing, had forgotten his Polish family and even the Polish language? And when at last we managed the complete identification of a child, new problems arose. What was to be done if the child no longer had any close relative in Poland?

'It was also claimed that it would be a shock to the child to be returned to its real family. That turned out to be false. Moreover, it constituted a pure and simple violation of international law. The authorities on whom it depended [the military authorities in the Western Allied occupation zones] took the view that it was preferable to leave the child in its present surroundings – in the interest of the child – instead of making it get used to new surroundings, unknown to the child. These were humanitarian explanations which covered essentially political motives.'

We asked how children who were identified reacted when they were told they were not German but Polish.

'Reactions varied. The younger ones, who remembered nothing, were surprised. They had to be prepared for the change. The others, the older ones, who constituted the majority of those we traced, between 30000 and 40000 of the 200000 who disappeared, accepted the situation with delight, particularly when we were able to tell them that their parents were still alive and waiting for them. This did not happen often. As you know, many parents were executed, either for resisting the invader or for other reasons. This is why the kidnapping of children for the purpose of Germanization was not discovered until after the war. We thought, the whole of Poland thought, that these children, like hundred of thousands of others, had been stolen to work for the Germans as slaves and that they had died in slavery.'

In June 1948 an American investigator of the International Refugee Organization went to see a retired police official named Freidrich Kubler. The American described the interview in his report. He told the man that the child he had been given by the Lebensborn was Polish. 'There's nothing Slavonic about the child,' the man replied. 'He's fair and has blue eyes. I shall give him up only if there is an anthropological report on his racial origins.' The inquiry took place in Munich in 1948. The racism evident in it is also present in documents or statements of more recent date.

This child had been traced thanks to an RuSHA document dated 10 October 1943 found by the Polish army at Lodz in 1945. It gave both the original names of forty-nine Polish children and the German names they had been given, as well as the addresses of the adoptive parents. Attached to the document was a slip giving the name and rank of the SS officer responsible for taking these children to Germany. He was Kurt Heinze, head of the Oberweiss Lebensborn home in Austria and the recognized transport officer of the 'L office'. For a time he acted as head of its statistical office, and then became head of the R4 department, which was in charge of the Germanization programmes. Heinze played the same key role in kidnapping that Ebner played in the selection of illegitimate Nordics. A married man with five children at Burghausen, he had adopted a little Polish girl who was given the German name of Christine Glaser, whom he had placed as a housemaid with Dr Tesch at Steinhöring.

Nothing happened to him after the war. He lived in a comfortable villa, and did not hesitate to commit perjury in court several times, claiming that, as he was born in Poznan, he knew 'all the German families' in the area. There were tens of thousands of them. The disconcerting ease with which this SS member of long standing was able to get away with it is easily explained: certain of the judges who dealt with these cases of the adoption of foreign children sided with those former Nazi families who in all circumstances wanted to keep their kidnapped children, for, like many university professors, teachers and lawyers, they also were ex-Nazis who had been left in their positions; so it was inconceivable that they should decide against their German fellow-citizens in cases involving Polish sub-humans who, to make matters worse, had now become Communists. There is no lack of examples of this post-war Nazi justice which decided the fate of innocent victims in favour of their tormentors.

Marian Gajewy was seven when he was kidnapped at Poznan and taken to Germany under the name of Marian Gawner. After passing through Kalisz and Oberweiss, he was handed over to the family of Karl Dengler at Esslingen in 1943. An important point for the judges who dealt with the case in 1948 was that Dengler had been a member of the SS in 1933. His number was 105646 and his rank was that of Rottenführer (corporal). In 1946 the child was traced by a

Polish investigator, who took the steps necessary to secure his repatriation through the International Refugee Organization. As was frequently the case, these took up a great deal of time. Meanwhile Dengler took action through the German courts to adopt the child legally. Actually all adoptions made during the Nazi period had been annulled, and post-war adoptions were forbidden without previous reference to the Allied authorities. But Dengler won his case. The adoption was formally registered on 22 April 1948, constituting an unusual event in German legal history, for Dengler was only forty-five and German law forbids adoption by anyone under the age of fifty.

The partiality of the judges reached its climax in July 1948. The Allied investigators had declared the child to be of 'indeterminate' nationality, but the former Rottenführer Dengler appealed to the former Untersturmführer Heinze, who declared under oath that the child was *Reichsdeutsch*, 'pure German'. Meanwhile Marian, who was then thirteen, received this letter from his aunt, his mother having died in 1941.

> We have been happy since the International Red Cross informed us that you are alive. Your brother Henio asks when you are coming. Please come as quickly as you can. I am your dead mother's older sister. Perhaps you remember my sons Dziunio and Rysio, and your uncle Stasieks, who is my husband. We should like to take the place of your parents. Keep well. Your aunt, Wera.

Meanwhile, the parish priest declared that Dengler, who worked for Kodak, was a fanatical SS man, and that his wife shared his views, and the Polish investigating committee had produced conclusive evidence of Marian's origin. In spite of all this Marian Gajewy stayed Karl Dengler's 'son', as a 'child left for the German economy'.

The case of Alojzy Twardecki, renamed Alfred Hartmann by the Lebensborn organization, who was taken in the middle of the night from his bed at Rogozno and was found four years later in the French Zone at Koblenz, is even more indicative of the spirit of justice that presided over the restoration of kidnapped children to their families.

In 1973, accompanied by Alojzy Twardecki, we retraced the route that had taken him from his home to a handsome villa overlooking Horchheim on the outskirts of Koblenz. Together we went through the documents that make up his voluminous dossier, questioned witnesses of his 'German' youth, met his mother and his cousin Leon, who was abducted at the same time as he (and whose story we have already told on pages 167–8). It was hard to believe that perjury and bad faith had been able to prevail over a mother's appeals and definite proof of identity. Frau Weber, who was then 'Alfred Hartmann's' French teacher, told us:

'One day in 1948 Herr Bindenberger came and told me that the child would not be coming to school for a while. He seemed very frightened, and told me that he had to hide him, because a Polish committee of inquiry was coming to collect him. He added that he would bring Alfred back as soon as the Poles had gone.'

The child was in fact hidden for weeks, until the Poles had left. But the men and women in charge of returning these children to their own country did not give up so easily, and handed over the matter to the French occupying authorities. Here is the first report of the French investigator, dated 13 August 1949:

I went to see Frau Bindenberger at Horchheim, district of Koblenz. Frau Bindenberger received me in a friendly but extremely detached fashion, and at first behaved as if the whole matter was absolutely no concern of hers. She does not see why I should take any interest in Alfred, whom she adopted in 1944 and intends to keep at all costs. I explained to her that I perfectly understood her attachment to the child, but that this was a Polish child who had been deported by the Germans, handed over to the Lebensborn, and was now claimed by its mother. Frau Bindenberger told me she is completely unable to understand how the child could be claimed by his mother, since she is quite certain that he is a German orphan. I gave her to understand that I was perfectly prepared to admit her good faith, but that all Lebensborn children had been handed over to families as German orphans, which was absolutely false.

She had applied to the Lebensborn at Munich which, after many inquiries, had authorized her to go to Polzin, near Berlin, in May 1944, to choose a child. This is what they told her there about the boy. His name was stated to be Alfred Hartmann, born on 24 March 1938 at Poznan, and his parents were dead; his father had been in the German army and had been killed in action. She claims that she saw a birth certificate, but has never seen a death certificate of the parents; everything was to have been settled by the Lebensborn after the war.

Gradually Frau Bindenberger began dropping her assured and slightly aggressive attitude and said that really she is very attached to the child, whom she regards as absolutely her own, that it would be very painful to part from him, and that the child would certainly suffer too. The Polish Red Cross has been to see her once, and that evening Frau Bindenberger had told the whole story to her husband in the child's presence; it was thus that young Alfred, who did not even know that he was not their son, learnt his true origin. It seems that he wept all night.

Frau Bindenberger told me he is an extremely proud boy, who would certainly suffer from a change in his situation, because he is very proud of belonging to a prosperous, well-dressed and well-fed family. He is considered the happiest and most spoilt boy in his class; also he works very well, and was top of his class at the end of the school year.

After some time I also saw Herr Bindenberger, who is less nervous than his wife and adopted a conciliatory and understanding attitude, but confirms that it would be very painful for him to be parted from the boy. Frau Bindenberger would like to know whether she could get in touch with the mother, for she cannot believe that she can be claiming her child after so many years.

I hardly saw Alfred, who was playing in the street; he refused to come up, knowing that a person in uniform was with his mother. He is a big, very robust boy, who scowled at me as I passed and refused to greet me.

The French investigator did not dare produce a letter from the boy's mother which had reached him some weeks before.

My dear son Alojzy,

At last we have found you. Have you forgotten your mother? We have looked for you everywhere. My dear Alojzy, Herr and Frau Bindenberger are not your father and mother, they are total strangers to you. Alojzy, don't you remember how the police took you away by force, can't you remember grandmother and your cousin Leon? He is already home, he was in Italy, and has not forgotten Polish. Alojzy, this is your mother writing to you, and I am at Rogozno, you were born here. Come home to your mother soon, we are all waiting for you. Alojzy, you are Polish and not of German blood. Tell Herr Bindenberger that you want to come home, when you receive this letter write a few lines. Your mother greets you and embraces you impatiently.

19 May 1949

Herr Bindenberger, a former aide of the Gauleiter of the Rhineland, immediately and without the investigator's knowledge took steps to have the boy adopted, and Kurt Heinze hastened to his assistance from distant Bavaria and made the following declaration under oath:

My name is Kurt Heinze, born 26 January 1901, living at 10/11 Siedlung Kümmernis, Burghausen (Upper Bavaria).

While I was active in the Lebensborn organization, Munich, I made the acquaintance in 1944 of the child Alfred Hartmann. This child is to all appearances the son of German parents. As he was an orphan from the Poznan area and had no papers, a birth certificate was made out for him in which Poznan was stated to be his birth-place.

Burghausen, 14 August 1948

Both the dates and his statement about the boy's parentage were false, and the reason for the absence of a birth certificate was invented. He himself had escorted from Kalisz to Bad Polzin the party of children in which Alojzy, alias Alfred, was included. None the less the adoption certificate was completed in the presence of the Koblenz notary Dr Kleinle on 13 January 1948. On page 5 this states that 'the child's mother is unknown'.

Why the child's mother and not his family, or father and mother? Obviously because it was perfectly well known that the boy was the son of Malgorzata Twardecka, who had been bombarding the Bindenbergers with letters for nearly two years.

On 14 February 1949 General Koenig, the French Commander-in-Chief in Germany, wrote to the head of the International Refugee Organization at Neuenburg a short letter in which he said he 'had the honour of enclosing a copy of the adoption papers of the child Hartmann-Twardecki, born 24 March 1938'. This was very strange, as the name of Twardecki does not appear anywhere in the adoption documents.

The years went by. After the death of his wife, Herr Bindenberger married again. Some time later he died. Alojzy was then fifteen. One day, after a quarrel with his new German 'mother', he decided to leave of his own accord. 'I'm going', he said to Peter Holler, his best friend, who still remembers the conversation. 'Going? Going where?' 'I'm going.' Alojzy had been brought up to hate the Poles, and was ashamed to say he had decided to go back to Poland. He was the last Polish child to be repatriated, in September 1953. This is his story:

'I had become a fanatical Nazi. I wept with rage when the men condemned at the Nuremberg trial were hanged. It took years before I stopped hating the Poles, as I had been taught, and then the French, who occupied our city of Koblenz. Actually it was a coincidence, for purely family reasons, that I became Polish again. My German mother was dead, and my Polish mother wanted me back. So one day I cleared off.'

Lutka Desca was not yet three when she was taken from her family at Lodz. The Lebensborn handed her over to the Bauer family at Stuttgart, and she reached the age of nine without knowing that she was Polish and that her father was still living in Lodz. The Bauers, like many other German families, had been careful not to declare that they had been given a child by the Lebensborn. But in 1949 they decided to divorce. The child, by order of the court, was entrusted to the care of the adoptive father, and to get her own back Frau Bauer bluntly told the little girl that she was not German but Polish and that she came from Lodz. After that the child started hating her 'father' and asked to be sent back to Poland.

In March 1947 two investigators of the Polish commission stopped for lunch at an inn at Esslingen in Baden-Württemberg. They were served by the innkeeper's son, a fair, blue-eyed boy of about eight. They engaged in an animated conversation, and the child hesitated as he placed the plates in front of them. They took no notice, but the same thing happened several times. The boy lingered near the table, listening intently and gazing at them.

'You understand Polish?' one of them asked, on the off-chance. The boy shook his head, but did not go away.

'You are the innkeeper's son, aren't you?'

The child hesitated, looked towards the kitchen, and nodded.

The other Pole then started murmuring the first words of a prayer that all Polish children without exception are taught almost as soon as they are able to talk, a prayer that only a Polish child could know. 'Angel of God, my guardian angel, stay always by my side; in the morning, during the day, and . . .'

'. . . and in the night, come always to my aid. Amen,' the small boy continued unhesitatingly, as if by reflex action. A few weeks later he was repatriated in spite of the opposition of the German couple who were on the point of officially adopting him.

Most of these adoptions took place in Germany between 1947 and 1950. We contacted the competent authorities in West Germany for information concerning these adoptions but they have no statistics for this period. This is unfortunate, as we were obliged to devote a great deal of time and trouble to the reconstitution of a situation that the post-war documents enable one to foresee. The favour shown by German courts to the adoptive parents led in most cases to the abandonment of the investigations conducted by the commissions of the countries concerned. Children Germanized under Hitler became German definitively and came legally and irreversibly under West German jurisdiction. The children most affected were those whose parents had been killed, because 'in the interest of the child' the adoptive parents were given preference over relatives of the second degree such as aunts, uncles or grandparents.

Ivan Petrochik, born in a small village in Yugoslavia on 1 January 1941, was kidnapped in 1943 by an SS detachment acting for the RuSHA. His father had been shot by the Gestapo and his mother sent to a concentration camp. The mother survived, however, and began searching for her son as soon as she was liberated. The search lasted for seven years, for it was only in 1952 that the child was traced to a family at Kassel who had adopted him in 1949, although they were well aware of his origin, for the Lebensborn file stated that he was the child of a Yugoslav 'bandit'. The Belgrade authorities intervened. Yugoslavia enjoyed a better reputation than Poland in the West because of its position in the Cold War, and consequently had greater freedom of action. Ivan's mother won her case and the adoption was officially annulled. That was in 1952.

Chancellor Adenauer had been in office for three years, and he was concerned, not only with his country's economic reconstruction, but with recovering its place in the civilized world. The Federal Republic, which had been a member of the Council of Europe for two years, was anxious to do everything possible to efface memories of the tragedies of the Nazi period. At this time it began the denazification trials, in connection with which certain obligations towards the West, on which it depended, had to be respected. Yugoslavia – and this is an important point – was then being subjected to a violent 'anti-revisionist' campaign by the Soviet Union, supported by Poland and Czechoslovakia. The principal beneficiaries of this political climate were the families of former members of the SS to whom children had been entrusted by the Lebensborn organization, with the result that many thousands remained in Germany. Some, whose traces had been lost even before the collapse of the Nazi régime, know nothing about their past. Others, officially adopted or too Germanized to start again, to return to a country they had been taught to despise, knew their past. Nevertheless they chose to remain. Some went so far as to jump from trucks taking them to transit camps before their departure for Poland. During this period of administrative chaos some appalling dramas took place, dramas that surpassed in intensity those of the kidnappings. The children were older now and more difficult to handle as they were able to understand what was happening.

There were some adoptive parents who realized that it was better for the child to go home to its brothers, sisters, fathers, mothers, uncles, aunts or grandparents. The documents show these to have been a minority. Most of them were childless couples, or couples that had lost an only son in the war and had concentrated all their love on the adopted child. There were also dupes of the Lebensborn who genuinely believed they had adopted a child of pure German stock. Couples who returned the child after the war – and many of these have maintained an affectionate relationship after its return to Poland and to this day – are in retrospect more to be pitied than blamed. We met several such couples who pay annual visits to 'their child', now married in Poznan, Lodz or Warsaw. A twist of fate is that some of these 'inferior Poles' are the sole official heirs of West Germans, a situation which the Lebensborn charitable organization certainly did not foresee.

A large number of Russian and Polish children carried off in the transport of the retreating Wehrmacht during the last year of the war ended up in the Moohren monastery in the Unter-Thingen area, where many died of neglect. We did not get very far with our inquiries here, because the possible witnesses seemed to be suffering from loss of memory; all they remembered was that the children were very young. Some of the survivors were traced, others had been 'secretly distributed'.

One case in the files in which the words 'kidnapped Russian children' and 'Wehrmacht' are associated is sufficiently remarkable to be told. A Russian boy born in 1940 on the Lithuanian border was carried off to East Prussia by the Wehrmacht in 1944. He was one of about a hundred children thus kidnapped by the Wehrmacht and not the SS and dispersed in various camps all over Germany. After the war they were cared for by one of the numerous religious charitable organizations that dealt with 'unaccompanied' refugee children, and this boy was sent to the United States. Fifteen years later, in 1961, he returned to Germany as an American soldier. He had in his possession a small photograph on the back of which was the name of a little girl whom he thought might be his sister. He applied to the International Red Cross Tracing Service at Arolsen. Long months of inquiry resulted in his tracking down his sister, now a married woman with children living in London. She too had been carried off by the German troops in retreat. The vital clue came from statements made by a nurse at Bad Polzin when she was interrogated. She remembered the arrival of five parties of children, or about sixty children altogether, of whom the Tracing Service had succeeded in finding only twenty-five. The American GI and the British housewife were added to the list.

The history of the school for German girls at Achern, in Baden, is more complete. It was a *Heimschule*, or State boarding school, housed in a building belonging to the SS and devoted to the education of future German mothers; it is still referred to in the neighbourhood as the *Hitlerschule*. The pupils were of two kinds: girls of German stock (many from South Tyrol), and girls from Eastern Europe suitable for Germanization who were to serve as future 'breeding' mothers. Our inquiries in the charming little town, halfway between Strasbourg and Baden-Baden, were directed to clearing up two points: (a) what had the Lebensborn in mind for these girls from the East whom the SS intended to take back when they reached the age of fifteen or sixteen, and (b) what had become of them after the war? We have already given the answer to (a) in an earlier chapter, but on reading the files we discovered that a number of girls had refused to be repatriated and had remained in the area. A full report by a French investigator of the International Refugee Organization, begun in March 1948, that is, three years after the end of the war – and three years is a long time in the life of a girl of ten – enabled us to form an idea of the climate that still prevailed in Germany at that time. Here are some extracts:

I went to see the priest, an elderly man, who received me kindly. He knew the house at Illenau, the *Hitlerschule*, as it was called, where children were sent for *Eindeutschung* [Germanization]. He said that at the beginning the girls came from South Tyrol; there were about 300 of them, all Catholic, who went regularly to mass and to confession. Later there were Polish and Russian girls . . .

He knows that some children were placed in the area, and that their names had been Germanized. But he does not know exactly where, nor does he know what has become of them . . . He told me that at Sasbach there was a Ukrainian priest who looked after all those that had stayed, and that he surely knew the children placed in the area. He had encouraged the big girls not to go home because of the Russians. He gave me the address of the Ukrainian priest, named Boijczujcz, who has an office on the second floor of the town hall. At first the latter was rather reticent, but gradually unbent. He knew perfectly well what I was after. He had been deported from the Ukraine himself. At first he had had to work in a factory, but then he had been permitted to exercise his ministry, though under constant Gestapo supervision . . .

As for the *Heimschule*, he knew that it existed. One day when he was in Strasbourg, visiting the cathedral, which had been converted into a kind of museum, he met an Alsatian priest (from a village quite near Strasbourg). It was the latter who had first told him about the 'kidnappings of Polish children' and about the school at Achern. It seems, according to this priest (whose name unfortunately he never knew), that some children were sent to Alsace and Lorraine and were placed in families there . . .

The priest had of course never been able to go there, but he had been shattered by what he was told by his Alsatian colleague, and what he said is certainly true. The children were to be placed n 100 per cent Nazi non-Catholic families.

[On 24 April 1948] I went to Fautenbach to see the Schindler family at 54 Dorfstrasse who had with them the little Alice Sosinger (real name Alicya Sosinka [see p. 163]). I was immediately told that she had been repatriated with three other girls.

I called on the Kornmayers at Oensbach. Frau K. began by receiving me angrily and in tears, telling me that the child had been repatriated in appalling conditions and was surely dead. It would have been better for her to have been left with them. I told her it would have been better to have left her in Poland in the first place. She calmed down a little, and I succeeded in having a long conversation with her. She has, though less so than Frau Schindler, that kind of friendliness mixed with servility that is so common among the Germans. But underneath her rather rude behaviour she is in fact much more honest and certainly much more anxious to help.

The girl Danuta Wutzow had come to them in the following circumstances. One of their daughters, a dressmaker, had been sent by the Arbeitsamt [Labour Office] to work at Illenau. She worked there for about two years and a half, from 1942 to 1945. There she met little Danuta, who was six. She was a rather thin, pale little girl, but she was pretty and had 'race'. The little girl begged her to take her home with her, and she obtained permission to do so, first for one day and another time for three days. The little girl cried every time she had to go back.

Finally Frau Kornmayer applied to keep her for good. Her daughter was slightly deformed and she assumed she would not get married, so she wanted to adopt the little girl. A long correspondence with the Lebensborn ensued. They had to provide medical certificates for the whole of their family, and inquiries were made about them. Her husband worked in the post office, and as such was a member of the party, but they were known to be practising Catholics, and at first their application was refused. All this took six months. Then they were told they could have the guardianship of the child, with an allowance of twenty-five

Reichsmarks a month, on condition that they signed a document committing themselves to handing her back at twenty-four hours' notice. Frau Kornmayer told me that the little girl was highly valued from the racial point of view, and that they wanted her back later for purposes of reproduction.

She refused this offer, preferring not to be paid. She was able to keep the little girl. She did not answer the Lebensborn letters from Munich, and the girl stayed with her, but she had always lived in the fear that they would take her back . . .

I went to see the gardener's family at the end of the village whose address was given me by the priest at Achern. In fact this is the Fruh family, mentioned by the Polish Red Cross (the child Helen Winkenauer lives with them). I first saw Frau Fruh, alone. She was very talkative and seemed to be hiding nothing. This is how she came to adopt Helen.

In the summer of 1943 [?] she had a great deal of work. A neighbour told her to ask for a girl at Illenau to help gather the raspberries. She did not do so, but two days later four or five little girls of about twelve arrived. She gave them bread and butter and fruit, and let them loose in the garden, where they gathered raspberries. She was then told she could keep one of them. She said she was not interested, because she needed real help. One fine day Helen arrived and said she wanted to stay, so she kept her.

Frau Fruh told me she had always left Helen free to choose what she wanted to do. She had not prevented her either from returning to Poland or going elsewhere . . .

I saw Helen. She is a big girl, bursting with health. At first she was very taciturn, and seemed to have little desire to talk to me. She relaxed a little when I told her I did not want to force her to go back to Poland if she did not want to, but that I should like her to tell me about Illenau and other children who were being looked for by their families. When I suggested making inquiries with a view to finding her uncle, her face lit up, and she thanked me heartily when I left. Her real name is Ilona Helena Wilkanowicz. She was born at Pabjanice on 28 March 1931. She was brought up in an orphanage there . . .

Some children from the Tyrol had been there at the same time as she. She had arrived from Poland with about twenty little girls. The whole orphanage at Pabjanice had been evacuated. In 1943 about three convoys of children arrived from Poland. Fräulein K. did not think there were more than a hundred altogether, perhaps fewer . . . The Polish children were lodged in a separate building. There were four dormitories, with six or seven beds in each. There were only girls. They were all dressed alike, in a navy-blue uniform with a big collar . . . They had to speak German, but they generally spoke Polish among themselves. There was no religious instruction, and a priest was not allowed to come and see them. They were not prevented from going to mass, but the chapel had been converted into a theatre, the children realized that everything connected with religion was mocked at, and so preferred not to ask permission to go. Sometimes they hid under their blankets to say their prayers and make the sign of the cross.

The report of this French investigator, though written more than twenty-five years before, served as a curiously accurate guide in our inquiries into what had been one of the most important SS homes for kidnapped children. For the most part the adoptive parents, the Nazi officials of the home, as well as some of the children who 'had been left in Germany in their own interest' and whose names were mentioned in the report, had not changed their address. Only the former Nazi mayor of Achern, now retired and exclusively devoted to his garden, was away that day. Karl Steidel, the official who had been in charge of placing children in Aryan families, and whose name appears as 'adoptive father' in documents signed by Dr Ebner, was not expecting us; he has a 'horror of foreigners who stick their noses into German affairs'. However, in gratitude to the French, who had 'permitted him to earn his living as a gardener' after the war at the former SS school, then occupied by the French army, he eventually agreed to receive us.

The tone of the conversation could hardly have been colder. Karl Steidel had 'returned to the Poles the child entrusted to him by the Lebensborn', and considered that he had thus 'paid his debt to society'. The small sitting-room was dominated by an enormous photograph of an Alsatian dog, which enabled us to break the ice. 'It was a marvellous dog,' he said. 'It had no equal in hunting down Communists in Heidelberg in 1933.'

'Were you in the SS?'

'No, I was in the SA. The shock troops. But I have been

denazified, as they say. I have nothing to reproach myself with and, above all, I have no regrets. Understand?'

With his Tyrolean hat worn to the cord and several days' growth of beard, Karl Steidel, in spite of his abrupt speech and the sharp way in which he said '*kapiert*?' ('understand?') at the end of every sentence, was no longer likely to frighten anyone. In other times perhaps, and according to Helen certainly. Finally we got around to talking about the children.

'Children were taken from orphanages, only from orphanages. They were brought here to Germany because they had German blood in their veins. The child I adopted had no one left in Poland; her grandfather was dead, her grandmother could not keep her. But because the mother was of German blood she was put in an orphanage and from there brought here to Germany. Anyone with a spark of honesty will confirm what I say. They were orphans, orphans of German blood.'

We saw no point in pressing him further. International Refugee Organization documents dated 1948 state that the 'orphan' adopted by Karl Steidel was restored to her family after being repatriated to Poland in March 1948. Her name was Helena Fice.

When we asked why little girls at the Achern school were branded on the wrist and neck and were given hormone injections to make them reach puberty more quickly, Steidel flared up.

'All that's rubbish,' he said. 'The girls learnt dressmaking and knitting. They did a great deal of sport, but in 1945 *alles war aus* – it was all up.'

20. Open Wounds

'Every finished job on which there is no possibility of going back will be a victory for Germany.'
HEINRICH HIMMLER

An interview at Bordeaux, dated July 1973:

'As a child, my name was Ingrid de Fouw, born on 31 July 1944. One day I went to tidy some papers in an attic and there I found out my name and my Christian name. Since then I have always wondered who my parents were. I knew I no longer had them, but still.... Up to the age of thirteen I was brought up by the public assistance at Bar-le-Duc, and then I was handed over to Mme C. I have always known I was an orphan.

'At the public assistance I was told that my parents were French, that they were arrested during the war by the Germans and then deported to a concentration camp – and it was there, I was told, that I was born on 31 July 1944. I also believed that my resister parents were tortured before they died... My name was then Irène de Fouw. One day, when I was fourteen, my foster-mother asked me to go to the attic to tidy some papers. There, in an old kitchen dresser, I found a packet of letters. They were not in envelopes. and all were in the same sloping handwriting used by educated persons. Chance made me read one of them. "Dear madam," it said, "I am Ingrid de Fouw's godmother – I knew her mother well – I should like to come and see her. I am a nurse, and I was at Lamorlaye when she was born."

'I promptly went to see my foster-mother and asked for an explanation. As usual, she told me she knew nothing, and that in any case my mother was a good-for-nothing who had "slept with a *boche*", and that I should be no better... I took no notice of her, because my relations with her were really bad (when I had a motor-scooter accident when I was fifteen she told me I'd have done better to end up under the lorry). But the letter I found in the attic worried me. I imagined my mother was Swedish, my father I didn't know. But I decided that when I was old enough I would find them.

'I got married at eighteen, and it was then that I really started making inquiries. At the public assistance offices, at the prefecture and at the high court I came up against a wall of silence. The answer was always the same. "We have nothing about you." As I had to look after my son who had just been born, I temporarily gave up my efforts.

'I resumed them in 1968. I had been told that my surname sounded German or Swedish, so I wrote to the German and Swedish consulates, but without success. Then the International Red Cross at Geneva put me in contact with the International Tracing Service at Arolsen, from whom I received a letter on 14 October 1970. That letter I shall never forget. A photograph of a little girl was enclosed. The little girl had a card in front of her chest with 'Ingrid de Fouw, born 31 July 1944' on it. The surname and the date of birth were the same as mine, only the Christian name was different. The letter asked me to examine the photograph carefully. "Do you recognize yourself in it?" it said. I sat up all that night, gazing at the photograph and weeping; I did not know whether it was I. With the mirror in one hand and the photograph in the other, I compared the two. The more I looked at my face in the mirror, the more I realized that the little girl with the card had the same eyebrows, the same forehead and the same eyes as I. Ingrid de Fouw and Irène de Fouw were the same child. I wrote to Arolsen to tell them I recognized myself.

'A year later, on 30 August 1971, I received another letter. This is what it said: "You were born on 31 July 1944 in a maternity home run by German troops at Lamorlaye, near Chantilly . . . You were evacuated to Germany in about August 1944 . . . then repatriated to France on 26 August 1946."

'When I read the name of Lamorlaye I realized that the woman who had written to my foster-mother had spoken the truth. I was flabbergasted; I don't know why, I thought

Lamorlaye was in Germany. The letter from Arolsen said that unfortunately all the documents that might have given information about my parents had been destroyed at the end of the war. "We have made all possible inquiries in Germany, but the results have all been negative," the letter said.

'I know that horrible sentence by heart. A few months later I went to Lamorlaye to see the place where I was born. There I was told that my father could only have been a German officer, for the maternity home there had been reserved for the children of German officers. I also discovered that my mother – presumably French or Flemish – was not married to my father. I could not believe what I was told. I, who believed I was born in a concentration camp of parents belonging to the resistance. I, who had learned to hate the Germans responsible for my parents' death, suddenly discovered I was the daughter of a German officer.

'But, after all, it is better to have a German father than no father at all. As long as I live I shall do all I can to find my parents.'

At the Lamorlaye town hall only three children born in the home were later acknowledged by their mothers when birth certificates were issued for them. The others, including Ingrid, are still searching for their parents.

In Munich and Hamburg and all the other big towns in Germany the Red Cross still regularly puts up posters with photographs and this moving text in the middle of the usual advertisements.

Who knows our parents and our origin? Who can tell us our names, where we come from, when we were born?

We are all aged over twenty and do not even know our real names, when we were born, or where we come from. In the war we lost contact with our families, but we were so young then that we could not say who we were. Not only do we suffer from not knowing where we come from, we also have all sorts of difficulties and disadvantages in life, because we only have provisional personal papers, with assumed names and estimated dates of birth. For many years the tracing service of the German Red Cross has been trying to identify us and find our families.

Who can help in this, and who thinks he knows our families? Please examine our photographs carefully.

Thirty years after the end of the war the wounds are still open. Children are still searching for their parents and parents for their children. And the files of the International Tracing Service at Arolsen show that few children whose father or mother had anything to do with the Lebensborn organization have succeeded in finding their parents; and, when they have succeeded in tracking down a mother, in most cases the latter has refused to acknowledge the child.

We discovered from the newspaper *Der Freiwillige*, the organ of former SS men published in the Federal Republic, that the SS has its own search service. The following is from the March 1959 issue:

> To meet requests for information about missing children from former Lebensborn homes the names and addresses of persons who worked in those homes are urgently required...
>
> Who knows Manfred Trantner, born 9 January 1940? Trantner was born in the Kurmark Lebensborn home and it is believed that he was eventually at the Steinhöring Lebensborn home.
>
> Information to SS Officers' Search Service, Sepp Küsters, 41 Weserstrasse, Essen.

Having established that Manfred Trantner had applied neither to the German Red Cross nor to the International Tracing Service, but that the former SS had its own search service, we immediately telephoned Herr Sepp Küsters, who politely told us it was none of our business.

Helena Wilkanowicz was one of the children who were traced after the war but for various reasons refused to be repatriated to their country of origin. She insisted on our interview taking place in the presence of her three children. 'They have got to know,' she said.

'I was kidnapped in Poland, at Pabjanice. Three SS men came into the room and put us up against the wall. There were about a hundred children altogether. They immediately picked out the fair children with blue eyes, seven altogether, including me, though I do not have a drop of German blood in my veins. I was twelve years old at the time. My father, who tried to stop my being taken away, was threatened by the soldiers. They even said he would be sent to a concentration camp. But I have no idea what happened later, because we were taken immediately to the children's reception camp at Brockau. In November 1943 we were brought here, to the SS school at Illenau, Achern.

'The unsuitable children were sent away from the school and liquidated. They threatened to send us to a concentration camp at every opportunity. Somehow I managed to survive. Perhaps it was because I'm blonde, I don't know.'

Why had she not gone back to Poland after the war?

'I was afraid. We were told so many things here about the Russians, the Communists, and the things they did. I was seventeen then, a girl, you understand? I was simply afraid. Besides, no one was waiting for me in Poland. In the meantime my father had disappeared'.

Did she feel German after so many years?

'No, not in the least. They still call me a dirty Pole [*Dreckpolack*]. It's dreadful that one can still hear such a thing here in Germany. But one stays a dirty Pole, just as one stays a dirty Jew.'

Did she intend visiting Poland?

'Yes, in a few years, as soon as we can afford it. But I'm not well, I still suffer from the after-effects of those years. I'm homesick, and that gets me down. I simply cannot find peace. I pace to and fro in my flat, run to the cemetery . . . As soon as I rest for a moment the whole thing starts all over again, just like thirty years ago.'

In the basement of the Lidice museum, not far from Prague, the lights go down. Visitors from all over the world are still. They have come to see a documentary made here in Lidice in June 1942 by a German army film unit which the Third Reich propaganda services had labelled a 'cultural film'. A young woman is working the projector.

Soldiers of the Wehrmacht or the SS – it is hard to tell which – laugh while pointing to a village in flames. Houses collapse one after the other. Fire gradually spreads to all the narrow streets, the church, the school, the town hall. Haylofts flare up like torches and throw white flashes into the hall. The first reel ends.

The young woman busies herself with the projector, without speaking.

The second part of the cultural document is devoted to the fate reserved for the people of the village. Men collapse in dozens under a crackling hail of bullets. The young woman stays silent and motionless by the projector, staring at the screen. The film ends on a gay note. Germans are seen taking an open-air meal on a site they have just finished levelling. The young woman puts away the reels and covers the projector with a black cloth. Before we leave, our guide tells us that she was one of the children who survived Lidice. Among the men shot down were her father and her uncles. One of the women rounded up to be sent to Ravensbrück was her mother. She did not come back.

Every time the film is shown she relives the nightmare of her childhood. She too was placed in a German family by the Lebensborn.

At Rogozno, near Poznan, Mrs Ewertowska lives alone. She has lived alone since that morning on 27 September 1943 when three men wearing long leather coats came to take away her daughter.

'She was there, sitting on her little pot. They dressed her and carried her away,' she said, pointing to the spot where this had taken place. The small room is poorly furnished, but spotlessly clean, though water has to be fetched from the fountain near the town hall in the square. On a table by the bed is a photograph of Eugenia, who became Irene Ewert in the Lebensborn files.

'When I came back from the station where the mothers had gathered to try and stop the train from leaving, I threw myself on to the bed and howled with grief. The neighbours came running, but they could not help me. I had thought of hiding her, or running away with her, but they took the mothers too if they caused trouble. Yes, it was very hard. At night I was afraid. I imagined she was dead. It was dreadful, taking a child from her mother . . . It was cruel, but what could I do?'

Pointing to the cupboard against the left side of the bed, she went on:

'The people here tried to console me, and said: "Buy flour, sugar and butter, so as to be able to bake her a fine cake when she comes back . . . " For five years I had no news. Then one day she was found with the help of the Red Cross, but the German family with whom she lived would not part with her. When the gentleman of the commission for the repatriation of our children went there, the German family locked themselves in the lavatory for hours. They hid my child again and again. She never came back.'

Meanwhile she had opened the cupboard.

'Look. Everything I bought to celebrate her return is still there, everything.'

The poverty revealed by the contents of the three shelves made the idea of the party to celebrate her daughter's return that the old woman had been nursing for thirty years seem even more absurd; her days and nights are still haunted by the memory of a little girl in a white dress with short hair who was taken early one morning in 1943 and never came back. 'I still never stop thinking of her,' she said. 'I'm old, and I can't afford to go to Germany. She still lives there with that German family at Flensburg.'

A few weeks later we met Irene.

Eugenia Ewertowska, whom Dr Tesch had renamed Irene, is now forty years old. In 1947, while the Poles were still looking for her, she was officially adopted by a family named Horn. She is now a wife and the mother of two children, and does not want to hear of her mother. To her 'that Polish woman', whom she strangely resembles, died long ago. 'I feel nothing for her,' she said. 'What do they want of me? Those people never stop pestering me. All I want is to be left in peace.'

To her mother we had seemed a heaven-sent link on which she set high hopes, and when we left Rogozno she had given us a message. In view of Irene's aggressive attitude we did not dare give it to her.

Dear Irene, I still think of you always. At least write me a few lines. Or have you forgotten your real mother? You are not German, you are Polish. You are not Protestant but Catholic, as you were baptized in church in Poland. You were not born in Germany, but here in Rogozno in 1935. Your German mother is lying to you. You are Polish, from Rogozno. Dear Irene, one day you will be sorry you did not write to your old mother. My dearest Irene, I can bear it no longer. At least write me a few words before I die. What are you doing there at Flensburg? What are your children like and who is your husband? I am already sixty-seven; I am not so young any more. When I die here one day you will yet weep there at Flensburg.

According to the latest Polish Government statistics, only about 15 per cent of the Polish children torn from their families to be Germanized have been repatriated.

Postscript

The trial of the Race and Settlement Head Office and the Lebensborn, known as Case no. 8, opened at Nuremberg, in the American Zone of Occupation, on 10 October 1947 and ended in March the following year. Among the defendants were Max Sollmann, Gregor Ebner, Gunther Tesch and Inge Viermetz. One of the principal characters was missing: their head, Heinrich Himmler, who had committed suicide on 23 May 1945.

'It is quite clear from the evidence,' the court decided, 'that the Lebensborn Society, which existed long before the war, was a welfare organization, and in the beginning a maternity home. From the start it cared for mothers, both married and unmarried, and children, both legitimate and illegitimate. The prosecution has failed to prove with the requisite certainty the participation of the Lebensborn, and the defendants connected therewith. in the kidnapping programme conducted by the Nazis. While the evidence has disclosed that thousands upon thousands of children were unquestionably kidnapped by other agencies or organizations and brought to Germany, the evidence has further disclosed that only a small percentage of the total number ever found their way into Lebensborn homes. And of this number only in isolated instances did the Lebensborn take children who had a living parent . . . Lebensborn did not participate in the selection and examination of foreign children.'

The defendants, Sollmann, Ebner and Tesch, were therefore found guilty only of having belonged to the criminal organization of the SS. On the other charges they were acquitted.

Bibliography

The Lebensborn records – at any rate those that escaped destruction – consist of nearly 200 files, each containing between 300 and 500 documents, either originals or copies. These files, apart from those recording the establishment of the organization, include the voluminous correspondence exchanged over a period of nearly ten years (1936–45) between the patients who passed through the maternity homes and the SS administration attached to Himmler's headquarters.

The letters of these 'Nordic' mothers have been wrongly dismissed by historians as unimportant and have consequently never been published, though they have the inestimable advantage of throwing an inside light on many aspects of this so-called SS charitable organization.

Most of the documentation – here rescued from oblivion for the first time – is to be found at the International Search Service (SIR) set up by the Allies at Arolsen in the German Federal Republic, which works – fortunately for the student – under the supervision of the International Red Cross. The invaluable aid of this institution, and of its Swiss director, M. de Cocatrix, in particular, is the more appreciated in that it was extremely unusual, not to say exceptional. We take the opportunity here of expressing our gratitude for the warmth of our reception and, above all, our appreciation of the institution's impartiality.

We are also indebted to the following: the Wiener Library, of London, who enabled us to trace a number of witnesses, ex-members of the SS, now retired into anonymity; the Centre for Contemporary Documentation, of Paris, whose indefatigable workers assembled complete documentation of the plundering of Jewish communities for the benefit of the Lebensborns; the

Library of Contemporary International Documentation of the university of Nanterre, which possesses a number of books and documents of inestimable value which are not to be found elsewhere; and the Polish Commission for Hitlerite Crimes, without whose aid we should not have found evidence of the facts stated in the second half of the book.

Without the aid of these institutions and the friendliness and courtesy of the witnesses and victims whom we met this book could not have been written.

Notes

The principal sources used in the preparation of this book appear only rarely in the now familiar bibliographies of books about the SS. Our researches off the beaten track of the history of the Second World War enabled us to discover a considerable amount of unpublished material. This provided essential material about the Lebensborn organization, both in Germany and in occupied Europe.

Records and Documents Consulted

International Search Service, Arolsen.

Polish Commission for Hitlerite Crimes, Warsaw, Kalisz, Katowice.

Polish Red Cross, Warsaw.

German Red Cross, Hamburg.

Lidice Museum, Lidice.

Film Library of the German Democratic Republic, Potsdam (Nazi propaganda films about race, the family, the Germanic heritage, the Jews, etc.).

Film Library, Warsaw.

Wiener Library, London.

Centre de documentation juive contemporaine, Paris.

Bibliothèque de la documentation internationale contemporaine, Nanterre University.

Files of the *Süddeutscher Zeitung*, Munich.

Files of the International Refugee Organization.

American trial at Nuremberg, Case No. VIII, 1947–8.

German trial of heads of Lebensborn organization, Munich, 1950.

Unesco report, *La situation des enfants polonais sous l'occupation allemande*, Trogen, Switzerland, July, 1948.

Report of West German Ministry of the Interior on the Lebensborn organization, Bonn, April 1955.

Tagebuch, Hans Frank, Warsaw.

La Condition humaine sous la domination nazie (Europe occidentale), speech by M. Edgar Faure, joint prosecutor at the Nuremberg trial.

Cahiers Pologne-Allemagne, Paris, 1963.

Third Reich school textbooks.

Documents, Paris, monthly review of German questions.

Principal Books Consulted

Josef Ackermann, *Himmler als Ideologe*, Göttingen, 1970.

François Bayle, *Croix gammée ou Caducée*, Paris, 1950.

Hedwig Conrad-Martius, *Utopien der Menschenzüchtigung*, Munich 1955.

J. S. Conway, *The Nazi Persecution of the Churches*, London.

Jacques Dalarue, *Histoire de la Gestapo*, Paris.

Walter Darré, *Neuadel aus Blut und Boden*, Munich, 1934.

Heinz Guderian, *Erinnerungen eines Soldaten*, Heidelberg, 1951.

Hans Gunther, *Rassenkunde des deutschen Volkes*, Munich, 1928; *Adel und Rasse*, Munich, 1927.

Ernst Hanfstaengel, *Zwischen Weissem und Braunem Haus*, Munich, 1962.

Helmut Heiber, *Lettres de et à Himmler*, Paris, 1969.

Adolf Hitler, *Mein Kampf*.

Heinz Höhne, *Der Orden unter dem Totenkopf – die Geschicte der SS*, Gütersloh, 1967.

Robert M. W. Kempner, *Le IIIe Reich en procès*, Paris, 1972.

Felix Kersten, *Totenkopf und Treue*, Hamburg, 1954.

Robert Koel, *German Resettlement and Population Policy*, 1939–45, Cambridge.

Eugen Kogen, *L'État SS*, Paris, 1970.

Czeslaw Madajczyk, *Die deutschen Besatzungspolitik in Polen 1939–1945*, Wiesbaden, 1967.

Roger Manvell and Heinrich Fraenkel, *Heinrich Himmler*, London, 1965.

Léon Poliakov, *The Aryan Myth*, London, 1974.

Gerald Reitlinger, *The SS, Alibi of a Nation*, London.

Kiryl Sosnowski, *The Tragedy of Children under Nazi Rule*, Warsaw, 1962.

Albert Speer, *Inside the Third Reich*, London, 1970.

George H. Stein, *The Waffen SS*, New York, 1966.

Dr Ternon and Dr Helmann, *Histoire de la médecine SS*, Paris, 1969.

Joseph Wulf, *L'Industrie de l'horreur*, Paris, 1970.

Index

Compiled by Gordon Robinson